AF334584

Thyroid Hormone Receptors

METHODS IN MOLECULAR BIOLOGY™

John M. Walker, SERIES EDITOR

METHODS IN MOLECULAR BIOLOGY™

Thyroid Hormone Receptors

Methods and Protocols

Edited by

Aria Baniahmad

Genetic Institute
Justus-Liebig University
Giessen, Germany

Humana Press ✳ Totowa, New Jersey

Library of Congress Cataloging-in-Publication Data

Thyroid hormone receptors: methods and protocols / edited by Aria Baniahmad.
    p. ; cm. -- (Methods in molecular biology; v. 202)
  Includes bibliographical references and index.
  ISBN 0-89603-995-1 (alk. paper)
  1. Thyroid hormones--Receptors--Research--Methodology. I. Baniahmad, Aria. II. Series.

QP572.T5 T565 2002
573.4'7--dc21

                                                                        2002020686

Preface

The thyroid hormone receptors (TRs) are important regulators of a large number of biological processes in animal and human physiology. They play a major role in brain development, in the development of hearing, in maturation of bone growth, in amphibian morphogenesis, in the regulation of metabolic rate, in the differentiation of the intestine, in the control of heart rate, and in development of vertebrate photoreceptors and color vision. Aberrant functions of TRs induce tremendous defects in these pathways, including the human disease of Resistance to Thyroid Hormone, which is a genetically autosomal dominant inherited syndrome that is caused by mutations in the gene encoding the TRβ. The TRs are hormone-controlled intracellular regulators and thus the analyses of their mechanisms of action will contribute to our understanding of their physiology on the molecular level.

The aim of *Thyroid Hormone Receptors: Methods and Protocols* is to cover the major area of TR research and to offer an overview, as well as practical experimental details, of the major methods in the field for both newcomers and experienced researchers. The book will cover the up-to-date methodology for the diverse topics of TR research. Hereby, an introduction to each topic precedes several methods associated with the analysis of the specific research area. The topics include the analysis of heart rate, amphibian morphogenesis, and transcriptional regulation by TRs in cell-free systems, as well as in living cells. Also, the isolation of TR-regulated protein complexes, the functional role of TRs in vertebrates, and the analysis of the oncogene v-*erbA* in blood cell differentiation are described in this book. Additionally included are target gene analysis in brain and liver by various approaches, among them being detailed methods for microarray chip analysis to identify TR-target genes in liver.

The large spectrum of methods described here includes the generation of knockout mice, transgenic mice, transgenic *Xenopus*, and the use of the *Xenopus* system for analyses on the molecular and chromatin levels. Furthermore, microarray chip analysis, *in situ* hybridization, the in vitro transcription system, chromatin immunoprecipitation, baculovirus expression system, and various DNA transfer methods for cells and cell lines are described. In addition, methodology is presented for the isolation of in vivo protein complexes involved in TR-mediated gene control, the analysis of regulated chromatin and its hormonal regulation, the use of retroviral systems and primary blood cell culture, and heart rate measurements. Furthermore, the generation of null-mutant, transgenic mice and of *Xenopus*, the identification of mutant TRs from patients with RTH syndrome, and the functional analyses of these TRs are described. All these methods are provided, as in other Humana Press Methods in Molecular Biology titles, in a richly detailed manner including a Notes section.

Thus, *Thyroid Hormone Receptors: Methods and Protocols* covers all major topics and includes all the important methods of up-to-date research on thyroid hormone receptor biology. TRs play a most diverse role in animal physiology, one that is correlated with a nearly incomparable spectrum of approaches. Therefore, the various TR functions are being analyzed in different systems for a variety of purposes by both scientists and medical doctors. Subsequently, each specific system and approach to analyze TR functionality is associated with a specific spectrum of various methods. Therefore, to fulfill the reader's expectation of a book covering all major TR research topics with their corresponding methodology, our book is structured with sections devoted to specific topics. Each chapter provides an overview of a specific topic and the system analyzed, discussion of its potential to gain new insights into TR functionality, and, in great detail, the associated spectrum of methods. Therefore, the particularity of this book is that each chapter describes several specific methods required to analyze TR function within a specific TR-regulated biological process.

Thyroid Hormone Receptors: Methods and Protocols is suitable for both basic and applied researchers, including physicians, pharmacologists, technicians, and pathophysiologists working in the field of molecular endocrinology and TR-action. Since thyroid hormone receptors are members of the large supergene-family of nuclear hormone receptors, the methodology provided here may be of great value also to the investigation of other members of these hormone-controlled regulators.

Aria Baniahmad

Contents

Contributors

DENISE AUBERT • *Laboratoire de Biologie Moléculaire et Cellulaire de l'Ecole Normale Supérieure de Lyon, Lyon, France*

ARIA BANIAHMAD • *Genetic Institute, Justus-Liebig-University, Giessen, Germany*

JUAN BERNAL • *Instituto de Investigaciones Biomédicas "Alberto Sols", CSIC-UAM, Madrid, Spain*

OLIVIER CHASSANDE • *Laboratoire de Biologie Moléculaire et Cellulaire de l'Ecole Normale Supérieure de Lyon, Lyon, France*

SASHKO DAMJANOVSKI • *Unit on Molecular Morphogenesis, Laboratory of Molecular Embryology, National Institute of Child Health and Human Development, National Institutes of Health, Bethesda, MD; Department of Zoology, University of Western Ontario, London, Ontario, Canada*

WOLFGANG H. DILLMANN • *Division of Endocrinology and Metabolism, University of California, San Diego, La Jolla, CA*

HELMUT DOTZLAW • *Genetic Institute, Justus-Liebig-University, Giessen, Germany*

XU FENG • *Molecular Regulation and Neuroendocrinology Section, Clinical Endocrinology Branch, National Institute of Diabetes and Digestive and Kidney Diseases, National Institutes of Health, Bethesda, MD*

FREDERIC FLAMANT • *Laboratoire de Biologie Moléculaire et Cellulaire de l'Ecole Normale Supérieure de Lyon, Lyon, France*

JOSEPH D. FONDELL • *Department of Physiology, University of Maryland School of Medicine, Baltimore, MD*

OLIVIER GANDRILLON • *Signalisations et Identités Cellulaires, Centre de Génétique Moléculaire et Cellulaire, UMR CNRS, Université Claude Bernard Lyon I, Villeurbanne Cedex, France*

KARINE GAUTHIER • *Laboratoire de Biologie Moléculaire et Cellulaire de l'Ecole Normale Supérieure de Lyon, Lyon, France*

BERND R. GLOSS • *Division of Endocrinology and Metabolism, University of California, San Diego, La Jolla, CA*

ANA GUADAÑO-FERRAZ • *Instituto de Investigaciones Biomédicas "Alberto Sols", CSIC-UAM, Madrid, Spain*

PAUL MELTZER • *Cancer Genetics Branch, National Human Genome Research Institute, National Institutes of Health, Bethesda, MD*

MARTIN L. PRIVALSKY • *Section of Microbiology, Division of Biological Sciences, University of California at Davis, Davis, CA*

LAURENT M. SACHS • *Laboratoire de Physiologie, MNHN, UMR CNRS, Paris, France*

JACQUES SAMARUT • *Laboratoire de Biologie Moléculaire et Cellulaire de l'Ecole Normale Supérieure de Lyon, Lyon, France*

YUN-BO SHI • *Unit on Molecular Morphogenesis, Laboratory of Molecular Embryology, National Insitute of Child Health and Human Development, National Institutes of Health, Bethesda, MD*

JIEMIN WONG • *Department of Molecular and Cellular Biology, Baylor College of Medicine, Houston, TX*

PAUL M. YEN • *Molecular Regulation and Neuroendocrinology Section, Clinical Endocrinology Branch, National Institute of Diabetes and Digestive and Kidney Diseases, National Institutes of Health, Bethesda, MD*

SUNNIE M. YOH • *Section of Microbiology, Division of Biological Sciences, University of California at Davis, Davis, CA*

Introduction to Thyroid Hormone Receptors

Aria Baniahmad

1. Biology of Thyroid Hormone Receptors

Thyroid hormone receptors (TRs) play a major role in animal physiology. TRs are important and very interesting regulators of diverse aspects, including brain development, hearing, bone growth, morphogenesis, metabolism, intestine, and heart rate in vertebrates (**Fig. 1**). Aberrant functions of TRs induce tremendous defects in these pathways. For example, the human disease of Resistance to Thyroid Hormone (RTH) (*see* Chapter 8 by Yoh and Privalsky) is a genetically autosomal dominant inherited syndrome that is caused by mutations in the gene encoding the TRβ. The role of the ligand of TRs, the thyroid hormone, is to modulate the activity and functionality of TRs.

Two separate genes encode two highly homologous TRs, TRα and TRβ. The *TRα* gene is localized on chromosome 17, while chromosome 3 harbors the TRβ locus. Each gene encodes for several isoforms due to alternative splicing and alternative promoter usage (**Fig. 2**) *(1–3)*. The expression patterns of TRα and TRβ are different, although overlapping in developing and adult tissue *(4–6)*. Also, splice variants and the various gene products from both TRα and *TRβ* gene loci, which are derived from alternative promoter usage, exhibit a distinct expression profile. Since these naturally occurring "truncated" receptors affect the functionality of the full-length TRs, the different expression levels and expression profiles result in differing tissue specificity of TR action and modulation of thyroid hormone response.

TRs were cloned based on their homologies to the v-erbA oncogene in the avian erythroblastosis retrovirus *(7,8)*. The retrovirus induces erythroleukemia and sarcomas (*see* Chapter 6 by Gandrillon, and references herein). The v-erbA oncogene represents a mutant form of the TRα. Mutations include nine point mutations in the hormone binding region that lead to changes of single amino

From: *Methods in Molecular Biology, Vol. 202: Thyroid Hormone Receptors: Methods and Protocols*
Edited by: A. Baniahmad © Humana Press Inc., Totowa, NJ

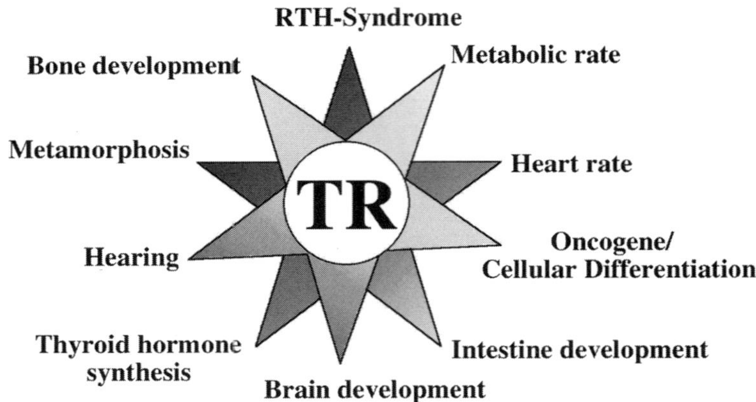

Fig. 1. The roles of TRs in the broad spectrum of animal physiology. The TR is a major regulator of vertebrate development involved in a great variety of different processes. Animal model systems including TR gene knock-out in mice and analyses of mutant TRs revealed important roles of TRs in the indicated vertebrate physiology. Mutations in the genes for TR lead to mutant receptors that induce diseases such as the RTH syndrome or functions as an oncogene, i.e., the v-erbA oncogene.

acids, two small deletions of a few amino acids in both the very amino-terminal and carboxyl-terminal receptor parts, and a gag fusion to the N-terminus (**Fig. 3**)*(9)*. The basis of the oncogeneity of the v-erbA oncogene is largely unknown, however, it is thought to be due to dominant negative action on thyroid hormone and retinoic acid receptor response *(10)*(*see* Chapter 6 by Gandrillon).

Based on the sequence similarities, structural motifs, and functionality, the TRs belong to the super family of nuclear hormone receptors (NHR). These receptors represent hormone-regulated transcription factors. Members of NHR include receptors for lipophilic hormones, such as steroids, receptor for non-steroids, and the orphan receptors, which are transcription factors with similar structures but no known ligand *(11)*. Receptors for steroids include the receptors for glucocorticoids, mineralocorticoid, progestins, androgens, and estrogens, whereas vitamin D, retinoids, prostaglandins, together with thyroid hormone, bind to receptors for nonsteroids. Although NHR regulate genes in a very similar manner, there are notable differences in the mechanism of action between steroid and nonsteroid receptors. In general, nonsteroid receptors prefer heterodimerization with the retinoid X receptor (RXR) and are bound to DNA in the absence of ligand. In contrast, the receptors for steroids, such as glucocorticoid receptor (GR) and androgen receptor (AR), are predominantly localized in the cytoplasm complexed with heat shock proteins in the absence of ligand. Cognate hormone

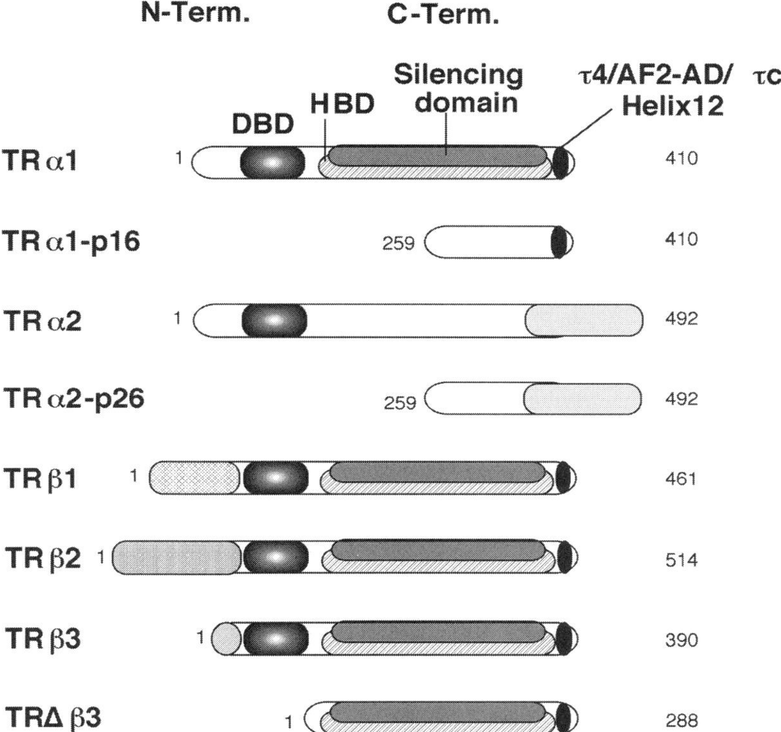

Fig. 2. Schematic view of the various TRs. TRα and TRβ are encoded by two different genes on different chromosomes. Each gene encodes for various subtypes of TRs due to alternative splicing or promoter usage. Indicated are the DBD, the hormone-binding domain (HBD), the silencing domain (active repression), and the helix 12 at the receptor carboxyterminus, which is essential for ligand-dependent transactivation. TRα2 cannot bind to thyroid hormone due to alternative splicing, which leads to a nonfunctional HBD. TRβ1, TRβ2, and TRβ3 contain different amino acid sequence in their amino termini. The numbers indicate the length of the receptor forms.

binding leads to a conformational change of the receptors, subsequent nuclear translocation, and gene activation. In general, binding of the hormone by non-steroid receptors also leads to a conformational change and to gene activation.

2. Thyroid Hormone

Thyroid hormone, isolated by Kendall in 1915, is one of the first hormones identified in the early last century. Its chemical structure has been known since 1925. Thyroid hormone is synthesized in the thyroid gland. It contains iodine atoms, and its synthesis is based on the amino acid tyrosine.

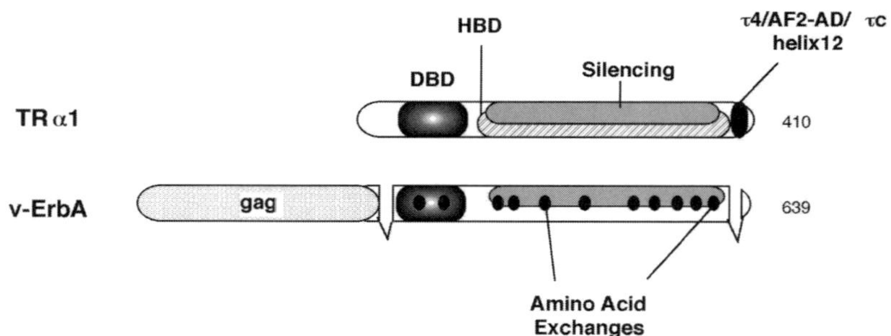

Fig. 3. Schematic view of differences between TRa1 and the oncogene v-erbA. The oncoprotein v-erbA lacks helix 12 and the first few amino acids compared to the wild-type TRα. Furthermore, amino-terminal gag-fusion and several point mutations (black circles) that lead to amino acid exchanges are indicated. The oncoprotein has severely reduced hormone binding affinity, while the silencing function is not affected by the mutations.

The production of thyroid hormone is controlled by thyroid-stimulating hormone (TSH) secreted by the pituitary. TSH secretion itself is under the control of thyrotropin-releasing hormone (TRH), which is secreted from the hypothalamus. The production of thyroid hormone is negatively regulated in a feedback mechanism. Thereby, thyroid hormone, through binding to its nuclear receptors TRα and TRβ, inhibits the genes coding for TSHα, TSHβ, and TRH. This regulation and the feedback mechanism is referred to as the hypothalamus-pituitary-thyroid axis (*see* Chapter 8 by Yoh and Privalsky and Chapter 1 by Gauthier et al).

Before the cloning of the receptors for T3, thyroid hormone was known to play a major role in various biological processes. Thyroid hormone influences a multiplicity of complex cellular functions with still largely unknown mechanisms. The hormone regulates developmental processes, such as the central nervous system and morphogenesis. It also regulates growth, metabolic rate, body temperature, and myocardial contractility.

The control of the central nervous system by thyroid hormone has been known for many years from analyzing hypothyroid rats. The absence of thyroid hormone during maturation of the central nervous system leads to irreversible mental retardation (*12–14*, and references therein). There is retarded development of the neurophil and Purkinje cells accompanied by diminished dendritic branching, elongation, and altered distribution of dendritic spines, delayed cell proliferation, and migration. Furthermore, deficiencies in myelination have been observed.

Also, there is a profound role of thyroid hormone on the development of amphibians (*see* Chapter 9 by Damjanovski et al.). The metamorphosis of tadpoles to adult frogs is under strict control of thyroid hormone. In early developmental stages, TRs are present, but thyroid hormone is not produced, indicating an important biological role for unliganded TRs. Experimentally induced lack of thyroid hormone prevents the metamorphosis resulting in giant tadpoles, while addition of hormone to young tadpoles leads to earlier metamorphosis and very small frogs. However, exactly how thyroid hormones induce metamorphosis is still largely unknown.

Taken together, a large number of questions regarding the basis and mechanisms of the biological effects of TRs remain open.

3. Transcriptional Control by TRs

The analysis of the transcriptional regulatory properties of TR is an exciting field. There are multiple levels of how the activity of TRs can be regulated in a cell. TRs have the interesting characteristic of silencing gene expression (active gene repression) in the absence of thyroid hormone (T3). Addition of T3 renders the receptor from a gene silencer to a gene activator. Thus, the hormone acts as a "molecular switch" controlling the repression and activation of target gene expression. All three transcription functions, silencing, hormone binding, and gene activation, are localized in the receptor carboxyterminus *(15)*(**Fig. 2**) (*see* Chapter 7 by Dotzlaw and Baniahmad). Lack of hormone binding capability with subsequent lack of target gene activation leads to deleterious defects in vertebrates. Interestingly, this general description of TR-mediated gene regulation is also modulated by the type of TR-binding sequence. Depending on its binding sites, TRs are also able to repress promoter activity even in the presence of T3, suggesting that the functional properties of TR are modulated by the mode of interaction of specific DNA sequences with the DNA-binding domain (DBD). These DNA elements are so called negative thyroid hormone response elements (nTREs). Through binding to these elements, TR no longer represses these target genes in the absence of thyroidhormone. The mechanisms of this opposite effect of hormone on TR lies presumably in the nature of the TRE. It is thought that the DNA sequence induces a specific eceptor conformation, which leads to binding of histone deacetylases even in the presence of ligand *(16)*. There are also other mechanisms by which TRs regulate gene expression. The inhibition of the proto-oncogenes JUN-FOS-mediated gene activation is one example by which hormone-bound TR is able to repress genes activated by this transcription factor heterodimer activator protein 1 (AP1). This inhibition does not involve the DNA-binding of TR *(17)*.

Thus, TRs regulate gene expression by various mechanisms, on the one hand as a DNA-bound transcription factor, and on the other hand through protein–protein interaction without direct binding to DNA.

Furthermore, the complexity of the TR regulatory network is enhanced by its dimerization properties. TRs bind to DNA either as homodimers or as a heterodimer with the RXR, another member of the NHR super family, which is regulated by retinoids. This indicates that thyroid hormone and retinoid acids may have some pathways and target genes in common. Thereby, direct repeats, inverted, or everted palindromes of the DNA-sequence AGGTCA are recognized and bound specifically by TRs *(18)*. TR binding sites (thyroid hormone response elements) are found in the close vicinity of the promoter as well as far upstream or downstream of the transcription start site of TR-target genes.

Both the gene silencing of target genes by DNA-bound TR in the absence of ligand and gene activation in the presence of thyroid hormone involves so-called cofactors *(19)*. Silencing is mediated by both binding to basal transcription factors at the promoter and by recruitment of histone deacetylase activity through binding to corepressors *(20)*. It is thought that nucleosome deacetylation leads to a more compact structure of chromatin, which exhibits lower accessibility for transcriptional activators and basal transcription factors. The modification and remodeling of chromatin involves large protein complexes that contain corepressors, and coactivators together with enzymatic activities for histone modification (*see* Chapter 10 by Wong).

Binding of thyroid hormone to the ligand-binding domain (LBD) of TR leads to conformational changes in the receptor C terminus *(21)*. Subsequently, corepressors are dissociated from the receptor, and coactivators are able to bind to the receptor C terminus in a hormone-dependent manner. The receptor with the associated coactivator complexes activates gene expression of target genes. There are two types of coactivator complexes: those which recruit histone acetylase activity, such as cAMP response element binding protein (CREB)-binding protein (CBP) or steroid receptor coactivator-1 (SRC1), and those which lack histone acetylase activity, such as the TRIP/DRIP-complex (*see* Chapter 11 by Fondell). Thus, the role of the hormone is to induce a conformational change of the receptor, which alters its transcriptional properties.

Both comparisons between the crystal structure and computer modeling from hormone-bound TRs (**Fig. 4**) and the closely related but unliganded RXR suggests that the major conformational change which is responsible for the hormone-induced receptor, is the helix 12 *(22,23)*. This helix has an important biological role for TR functionality. Upon hormone binding, the helix 12 is essential to relief silencing and to activate genes by inducing the dissociation of corepressors from the receptor and permitting the binding of coactivators *(24)*.

4. Diseases and Developmental Role of TRs

Several human diseases, including the syndrome of RTH (*see* Chapter 8 by Yoh and Privalsky), are based on malfunctioning of NHR helix 12. Upon hor-

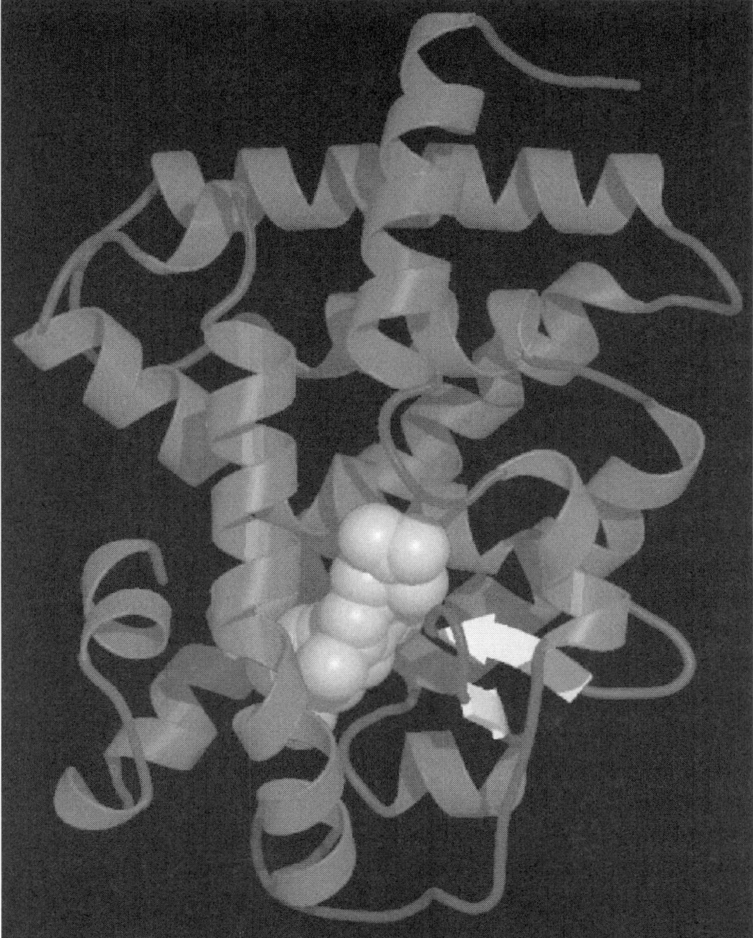

Fig. 4. Crystal structure of the TR HBD with the bound thyroid hormone. Crystal structure of liganded TRα HBD shows a predominantly α-helical structure with the pocket to bind thyroid hormone. The only two β-sheets are indicated as arrows. Kindly provided by R. Huber and R. J. Fletterick.

mone binding the helix 12 closes the hormone-binding cavity and is responsible for both corepressor dissociation and coactivator binding *(19,24)*. Mutations in the gene encoding TRβ, derived from patients with RTH, result in a complete loss or weakening of corepressor dissociation, despite the presence of hormone. Thus, it is expected that TR target genes regulated by classical TREs in patients with RTH are much more weakly activated or even strongly repressed despite the presence or even elevated levels of thyroid hormone. On the other hand, TR target genes regulated through negative TREs or through AP1 are more active compared to the normal situation.

Mutations in the *TRβ* gene, isolated from patients with RTH, are not only localized in the coding region of helix 12, rather there are three clusters within the hormone binding domain. These mutations affect dimerization function of TR with RXR, the inhibition of AP1 by TR, and lead mostly to reduced hormone binding affinity of TR. Similarly, the v-erbA oncogene product, a mutant form of TRα, lacks helix 12 (**Fig. 3**) and is, therefore, unable to dissociate corepressors. Thus, the oncoprotein exhibits a constitutive silencing function despite the presence of thyroid hormone *(24)*. It is believed that the oncogene product silences yet unknown genes that are important for cellular differentiation.

The role of TRs in development is being analyzed by characterization of patients with the syndrome of RTH, generation TR knock-out, and transgenic mice, as well as in the *Xenopus* system.

The phenotype of patients with RTH syndrome includes the symptoms of elevated levels of circulating thyroid hormone and decreased response to thyroid hormone. Various degrees of attention deficit, learning disabilities and mental retardation, hearing loss, and delay in bone growth and, therefore, short stature have been reported *(25)* (*see* Chapter 8 by Yoh and Privalsky). However, the precise role of TRβ inducing these symptoms is unknown. Interestingly, so far there is no human inherited disease described that is correlated with mutations in the gene encoding TRα. Mice model systems using knock-out of TRα or TRβ reveal distinct roles of these receptors in animal physiology *(26)*, (*see* Chapter 2 by Gauthier et al.). TRα is important for early development, including bone growth, maturation of the intestine, and proper development of the immune system *(27)*. Also, body temperature and heart rate is controlled by TRα *(28)*. TRβ, on the other hand, is involved in the maturation of cochlea, liver metabolism, and affects temperature control *(29)*. Furthermore, it was found recently that TRα2 null mutant mice exhibit loss of M-cones, which develop into green cone photoreceptors of the retina, indicating an association of *TRβ2* gene mutation with human cone disorders *(30)*.

Generation of mice carrying a mutation in the gene of TRβ, which is unable to bind to thyroid hormone, revealed severe abnormalities in cerebrellar development and learning *(31)*. This indicates a deleterious role of constitutive silencing function and corepressor association to unliganded TR in the brain. The role of corepressor association with TRs is approached by the analysis of transgenic mice expressing a dominant negative mutant corepressor (NcoRi) isoform in liver (*see* Chapter 3 by Feng et al.). Transgenic mice were generated that express in heart a mutant TRβ harboring a mutation identified in patients with the RTH-syndrome. These mice revealed that cardiac gene expression, prolonged cardiac muscle contractility, and electrocardiogram are comparable with a hypothyroid cardiac phenotype despite normal T3 levels (*see* Chapter 4 by Dillmann and Gloss).

Taken together, TRs are very important for a variety of different developmental aspects in vertebrates, including morphogenesis in amphibians and proper maturation of brain, bone, intestine, cochlea, green cone photoreceptors, metabolic rate, and heart rate.

5. Outlook

Research on TR is a very interesting and important field, which will provide exciting new information in the future. To shed light into mechanisms of how TRs exert their effects, the identification of TR target genes (genomics) is very important. Although a few TR target genes are known (*see* Chapter 5 by Bernal and Guadaño-Ferrez), at the present stage, only little is known about the identity of genes regulated by TR. It still remains unclear which dysregulated genes are responsible for mental retardation, hearing disorders, bone growth, heart rate (*see* Chapter 4 by Dillmann and Gloss), morphogenesis (*see* Chapter 9 by Damjanovski et al.) and the induction of cancer by the oncogene product v-erbA (*see* Chapter 6 by Gandrillon). Also, further analyses need to be performed to analyze the cellular networking of TR in the context of other cellular factors, coregulators, and chromatin (*see* Chapter 10 by Wong and and Chapter 11 by Fondell), as well as the mechanisms of cross talk in the various and highly specialized tissues. The detailed mechanisms of tissue response to TRs, in the absence of ligand and presence of thyroid hormone, require further characterization, e.g., at the level of proteomics.

In addition, TRs may not only exert their regulatory roles at the genomic and transcriptional level, but also at the nongenomic level *(32)*. These nongenomic activities of TRs may take place in the cytoplasm, although formerly, TRs were generally thought to be exclusively localized in the cell nucleus in both the absence and presence of thyroid hormone. Therefore, intracellular transport and phosphorylation events are considered to be involved in TR functionality *(33,34)* for which the mechanisms need to be elucidated.

Based on the high evolutionarily conservation of TRs, the generation of mice model systems will provide new important information about the role of each of the TRs in tissues and animal physiology. In combination with the analysis of transgenic mice and knock-in mice, introducing specific mutations in TR genes will provide mice model systems for human diseases. This approach will permit the identification of dysregulated target genes that cause specific symptoms.

Thus, because TRs possess broad effects in animal physiology with a broad spectrum of networking in cells, it requires the analysis of TR functionality at multiple levels: in the animal systems, in cell culture, and in vitro.

Based on the overall similarities (structural, biochemical, and functional) among receptors for nonsteroids and most orphan receptors, the methods described in this book may be also applicable to other members of the NHR super family.

Taken together, the TRs play multiple roles in a variety of different biological aspects in vertebrates. Brain development, hearing, bone growth, morphogenesis, metabolic rate, and myocardial contractility are the major known biological roles of TR, and gene silencing and activation are the major known functions of TRs and thyroid hormone. Therefore, the functional and biochemical roles of TR are being analyzed using different biological systems. Each system requires a spectrum of methodology. This book covers the major area of TR research divided into several chapters, each chapter covering one topic. Thus, each chapter describes not only one but a set of different methods required for analysis of TR research in a specific topic.

References

1. Chassande, O., Fraichard, A., Gauthier, K., et al. (1997) Identification of transcripts initiated from an internal promoter in the c-erbA alpha locus that encode inhibitors of retinoic acid receptor-alpha and triiodothyronine receptor activities. *Mol. Endocrinol.* **11,**1278–1290.
2. Munoz, A. and Bernal, J. (1997) Biological activities of thyroid hormone receptors. *Eur. J. Endocrinol.* **137,** 433–445.
3. Williams, G. R. (2000) Cloning and characterization of two novel thyroid hormone receptor β isoforms. *Mol. Cell Biol.* **20,** 8329–8342.
4. Forrest, D., Sjoberg, M., and Vennstrom, B.(1990) Contrasting developmental and tissue-specific expression of alpha and beta thyroid hormone receptor genes. *EMBO J.* **9,** 1519–1528.
5. Strait, K. A., Schwartz, H. L., Perez-Castillo, A., and Oppenheimer, J. H. (1990) Relationship of c-erbA mRNA content to tissue triiodothyronine nuclear binding capacity and function in developing and adult rats. *J. Biol. Chem.* **265,** 10,514–10,521.
6. Bradley, D. J., Towle, H. C., and Young, W. S., 3rd. (1994) Alpha and beta thyroid hormone receptor (TR) gene expression during auditory neurogenesis: evidence for TR isoform-specific transcriptional regulation in vivo. *Proc. Natl. Acad. Sci. USA* **91,** 439–443.
7. Sap, J., Munoz, A., Damm, K., et al. (1986) The c-erb-A protein is a high-affinity receptor for thyroid hormone. *Nature* **324,** 635–664.
8. Weinberger, C., Thompson, C. C., Ong, E. S., Lebo, R., Gruol, D. J., and Evans, R. M. (1986) The c-erb-A gene encodes a thyroid hormone receptor. *Nature* **324,** 641–646.
9. Forrest, D., Munoz, A., Raynoschek, C., Vennstrom, B., and Beug, H. (1990) Requirement for the C-terminal domain of the v-erbA oncogene protein for biological function and transcriptional repression. *Oncogene* **5,** 309–316.

10. Thormeyer, D. and Baniahmad, A. (1999) The v-erbA oncogene (review). *Int. J. Mol. Med.* **4,** 351–358.

11. Mangelsdorf, D. J., Thummel, C., Beato, M., et al. (1995) The nuclear receptor superfamily: the second decade. *Cell* **83,** 835–839.

12. Dussault, J. H. and Ruel, J. (1987) Thyroid hormones and brain development. *Annu. Rev. Physiol.* **49,** 321–334.

13. Koibuchi, N. and Chin, W. W. (1998) ROR alpha gene expression in the perinatal rat cerebellum: ontogeny and thyroid hormone regulation. *Endocrinology* **139,** 2335–2341.

14. Chan, S. and Kilby, M. D. (2000) Thyroid hormone and central nervous system development. *J. Endocrinol.* **165,** 1–8.

15. Baniahmad, A., Kohne, A. C., and Renkawitz, R. (1992) A transferable silencing domain is present in the thyroid hormone receptor, in the v-erbA oncogene product and in the retinoic acid receptor. *EMBO J.* **11,** 1015–1023.

16. Sasaki, S., Lesoon-Wood, L. A., Dey, A., et al. (1999) Ligand-induced recruitment of a histone deacetylase in the negative-feedback regulation of the thyrotropin beta gene. *EMBO J.* **18,** 5389–5398.

17. Lopez, G., Schaufele, F., Webb, P., Holloway, J. M., Baxter, J. D., and Kushner, P. J. (1993) Positive and negative modulation of Jun action by thyroid hormone receptor at a unique AP1 site. *Mol. Cell Biol.* **13,** 3042–3049.

18. Keda, M., Rhee, M., and Chin, W. W. (1994) Thyroid hormone receptor monomer, homodimer, and heterodimer (with retinoid-X receptor) contact different nucleotide sequences in thyroid hormone response elements. *Endocrinology* **135,** 1628–1638.

19. Xu, L., Glass, C. K., and Rosenfeld, M. G. (1999) Coactivator and corepressor complexes in nuclear receptor function. *Curr. Opin. Genet. Dev.* **9,** 140–147.

20. Burke, L. J. and Baniahmad, A. (2000) Co-repressors 2000. *FASEB J.* **14,** 1876–1888.

21. Leng, X., Tsai, S. Y., O'Malley, B. W., and Tsai, M. J. (1993) Ligand-dependent conformational changes in thyroid hormone and retinoic acid receptors are potentially enhanced by heterodimerization with retinoic X receptor. *J. Steroid Biochem. Mol. Biol.* **46,** 643–661.

22. Wagner, R. L., Apriletti, J. W., McGrath, M. E., West, B. L., Baxter, J. D., and Fletterick, R. J. (1995) A structural role for hormone in the thyroid hormone receptor. *Nature* **78,** 690–697.

23. Egea, P. F., Klaholz, B. P., and Moras, D. (2000) Ligand-protein interactions in nuclear receptors of hormones. *FEBS Lett.* **476,** 62–67.

24. Baniahmad, A., Leng, X., Burris, T. P., Tsai, S. Y., Tsai, M. J., and O'Malley, B. W. (1995) The tau 4 activation domain of the thyroid hormone receptor is required for release of a putative corepressor(s) necessary for transcriptional silencing. *Mol. Cell Biol.* **15,** 76–86.

25. Refetoff, S. (1997) Resistance to thyroid hormone. *Curr. Ther. Endocrinol. Metab.* **6,** 132–134.

26. Forrest, D. and Vennstrom, B. (2000) Functions of thyroid hormone receptors in mice. *Thyroid* **10,** 41–52.

27. Arpin, C., Pihlgren, M., Fraichard, A., et al. (2000) Effects of T3Rα1 and T3Rα2 gene deletion on T and B lymphocyte development. *J. Immunol.* **164,** 152–160.
28. Wikstrom, L., Johansson, C., Salto, C., et al. (1998) Abnormal heart rate and body temperature in mice lacking thyroid hormone receptor alpha 1. *EMBO J.* **17,** 455–461.
29. Johansson, C., Gothe, S., Forrest, D., Vennstrom, B., and Thoren, P. (1999) Cardiovascular phenotype and temperature control in mice lacking thyroid hormone receptor-beta or both alpha1 and beta. *Am. J. Physiol.* **276,** H2006–H2012.
30. Ng, L., Hurley, J. B., Dierks, B., et al. A thyroid hormone receptor that is required for the development of green cone photoreceptors. *Nature* **27,** 94–98.
31. Hashimoto, K., Curty, F. H., Borges, P. P., et al. (2001) An unliganded thyroid hormone receptor causes severe neurological dysfunction. *Proc. Natl. Acad. Sci. USA* **98,** 3998–4003.
32. Simoncini, T., Hafezi-Moghadam, A., Brazil, D. P., Ley, K., Chin, W. W., and Liao, J. K. (2000) Interaction of oestrogen receptor with the regulatory subunit of phosphatidylinositol-3-OH kinase. *Nature* **407,** 538–541.
33. Baumann, C. T., Maruvada, P., Hager, G. L., and Yen, P. M. (2001) Nuclear cytoplasmic shuttling by thyroid hormone receptors. multiple protein interactions are required for nuclear retention. *J. Biol. Chem.* **276,** 11,237–11,245.
34. Hong, S. H. and Privalsky, M. L. (2000) The SMRT corepressor is regulated by a MEK-1 kinase pathway: inhibition of corepressor function is associated with SMRT phosphorylation and nuclear export. *Mol. Cell Biol.* **20,** 6612–6625.

2

Null Mutant Mice for Thyroid Hormone Receptors

**Karine Gauthier, Denise Aubert, Olivier Chassande,
Frederic Flamant, and Jacques Samarut**

1. Introduction

In mammals, thyroid hormones (TH) have been shown to control the post-natal development of many organs, such as brain, intestine and long bone, and to participate in the maintenance of homeostasis in adults by controlling basal metabolism, heart rate, and body temperature *(1,2)*. To ensure this last role, circulating TH concentrations are maintained very stable by a tight control of TH production. Indeed, TH, which is primarily synthesized in the thyroid gland, represses the production of two peptidic hormones, thyrotropin-releasing hormone (TRH) in the hypothalamus and thyroid-stimulating hormone (TSH) in the pituitary. TRH normally stimulates the production of TSH, which, in turn, stimulates the thyroid gland and thus permits an efficient TH production *(3)*.

TH are lipophilic molecules able to passively cross the membranes and bind to nuclear receptors, thyroid hormone receptors (TRs), which are transcription factors whose activity is modulated by ligand binding *(4)*. Four TRs have been described to date, TRα1, TRβ1, TRβ2 *(5)*, and TRβ3 *(6)*, encoded by two distinct loci *TRα* and *TRβ*. In addition three other isoforms are generated from the *TRα* locus, TRΔα2, TRΔα1, and TRΔα2, which do not bind TH and act in vitro as inhibitors of TR. Little is known on the mechanisms of action of TH in vivo, and even less is known about the specific roles played by each TR isotype or isoform in the transmission of TH signal. To address this question, different knock-out mice, in which the expression of one or more of the TRs is selectively abrogated, were generated by homologous recombination *(7–13)*. A number of different alleles of the *TRα* locus have been generated in an attempt to better understand the role of the different isoforms. The comparative phenotypic analyses of these different mutant strains allowed to conclude that:

From: *Methods in Molecular Biology, Vol. 202: Thyroid Hormone Receptors: Methods and Protocols*
Edited by: A. Baniahmad © Humana Press Inc., Totowa, NJ

- TRα1 is the main receptor implicated in the transduction of TH signal during postnatal development, and particularly in the control of body growth, maturation of intestine and bone, and development of the immune system *(9,10,14,15)*.
- TRβ is the main receptor involved in the maturation of cochlea *(16)*, and in the regulation of liver metabolism *(17)*.
- TRα1 cooperates with, respectively, TRβ2 to negatively control TSH production in the pituitary *(10,11,18,19)* and with TRβ1 to regulate body temperature and heart rate *(13,20)*.

Knock-out is now a widely used technique based on homologous recombination (HR) performed in embryonic stem (ES) cells. These cells are pluripotent cells from the inner cell mass of E 3.5 blastocysts, able to grow in culture, and to participate to the development of the embryo when injected into a host blastocyst.

To specifically delete a gene (here one of the TRs), a recombination vector has to be introduced into ES cells in culture. This vector contains two arms of homology corresponding to the surrounding genomic regions of the region to be deleted and is separated by a positive selection cassette that encodes a protein conferring cell resistance to a toxic drug. This cassette allows one to identify cells in which the plasmid has been integrated. Since homologous recombination is a rare event, most of the clones, isolated after selection, are the result of a random integration in the genome. A specific screening of resistant cells is thus performed to identify the cells harboring the deletion of one allele of the targeted locus and its replacement by the selection cassette. These cells are then injected into host blastocysts, which are in turn, reimplanted in pseudopregnant females. These females give birth to some chimeric mice containing a mixture of grafted and host cells. These chimera are then crossed with wild-type mice. If a germinal transmission of the mutation occurs, some of the pups are heterozygous for the mutation in each cell of the whole body. Further crosses between heterozygous animals give rise to mice homozygous for the mutation. All these steps are summarized in **Fig. 1**.

In this chapter, we will describe how to perform a knock-out starting from the construction of the targeting vector to obtain ES cell clones harboring the mutation on one allele. Neither the production of mutant mice from these ES cells nor the different methods used to analyze these mutant strains will be discussed here.

The method developed here is the most classical one aimed at mutating a specific locus in all tissues of the mouse with the mutation occurring as early as the fertilized oocyte stage. New developments of this technique, using the Cre-*lox*P system, provide us with the possibility to perform the mutation in a time- and tissue-specific manner during mouse development. This system will only be described in the Notes section (*see* **Notes 1 and 5**).

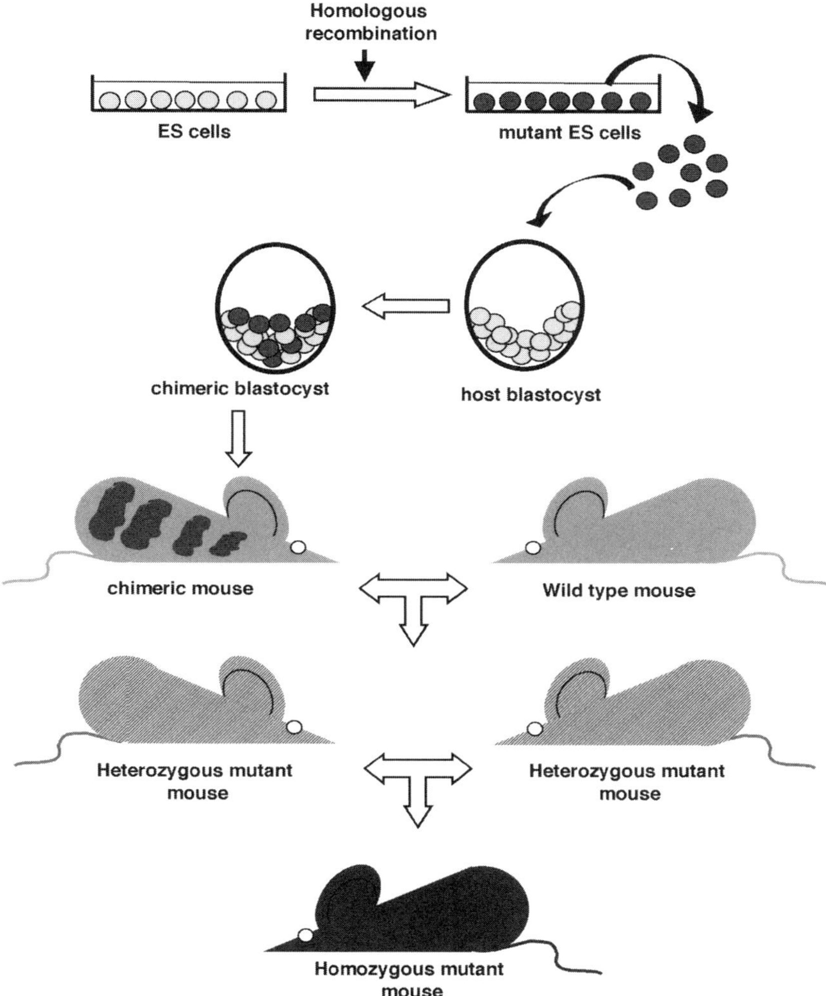

Fig. 1. Homologous recombination: from ES cells to mutant mice. Normal ES cells (figured light gray) are electroporated with the HR vector and selected for its integration at the right locus (cells in dark gray). Cells harboring the mutation on one of the two alleles of the targeted locus are injected into E 2.5 host blastocysts, which are then reimplanted in pseudopregnant females. The injected mutant ES cells participate to the development of the embryo, giving birth to chimeric mice composed of mutant and normal cells. Since ES cells and host blastocysts belong to two mouse strains recognizable by their hair color, the chimera can be easily identified. These chimera, usually males (because the ES cells used have a male genotype), are then crossed with wild-type females. Heterozygous animals are obtained and identified from their hair color. Heterozygous mice are intercrossed to generate homozygous animals in a mendelian ratio (1/4), provided that the mutation is not deleterious for embryonic development.

2. Materials

2.1. Construction of the Homologous Recombination Vector

1. Total genomic DNA from ES cells or a plasmid containing a fragment of genomic DNA covering the desired region.
2. Specific oligonucleotides designed to amplify the different arms of the homologous recombination vector.
3. A plasmid (pMC1Neo, Stratagene) containing the NeoR cDNA under the control of a promoter active in ES cells (e.g., phosphoglycerokinase [PGK]).
4. (Optional) A plasmid containing the herpes simplex thymidine kinase cDNA under the control of a promoter expressed in ES cells (e.g., PGK).
5. A PCR cloning kit (PGEMt, Promega; or Topo, Invitrogen; etc).
6. A *Taq* DNA polymerase able to amplify long DNA fragments with high fidelity (e.g., Expand Long Template, Roche).
7. 3 *M* Sodium acetate in ultrapure water. Store at room temperature (RT).
8. Ethanol 100%.
9. The restriction enzymes appropriate for the different cloning steps.

2.2. Homologous Recombination in ES Cells

Every material has to be sterile and tested for cell culture.

1. Fetal calf serum (FCS) tested for toxicity and cloning efficiency on ES cells.
2. Mouse embryonic fibroblasts (MEF) resistant to the antibiotic used for the positive selection (Gibco).
3. An ES cell line. We use ENS ES cells *(10)*.
4. Gelatin solution: 0.1% (w/v) tissue-culture grade gelatin mixed in ultrapure water and sterilized by autoclave. Store at RT.
5. Standard culture medium: Glasgow-modified essential medium (GMEM). Store at 4°C.
6. Penicillin–Streptomycin (PS): stock solution 100X, 10 g/L. Store at –20°C.
7. Glutamine (G): stock solution 100X, 200 m*M*. Store at –20°C.
8. Sodium pyruvate (NaP): stock solution 100X, 7.5% NaP. Store at 4°C.
9. Nonessential Amino Acids (NEAA). Stock solution 100X. Store at 4°C.
10. β-Mercaptoethanol : Stock solution 1000X, 10^{-1} *M* β-Mercaptoethanol in phosphate-buffered saline (PBS). Store at –20°C.
11. PBS without Ca^{2+} and Mg^{2+}. Store at room RT.
12. Mouse ESGRO^{TM} LIF 10^6 μ/mL (Gibco-BRL): stock solution 1000X. Alternatively, supernatant from transfected COS7 cells expressing the human recombinant leukemia-inhibitory factor (LIF), sterilized by filtration (0.22 μm). The amount of supernatant required has to be evaluated. Store at –20°C.
13. ES medium: GMEM, FCS 10%, PS 1X, G 1X, NaP 1X, NEAA 1X, β-Mercaptoethanol 1X, LIF 1X. Store at 4°C for a maximum of 15 d.
14. Freezing medium (2X): 80% (v/v) FCS, 20% (v/v) dimethyl sulfoxide (DMSO). Extemporaneously prepared.

15. Trypsin solution for ES cells (TES): 70% (w/v) NaCl, 10% (w/v) D-glucose, 3% (w/v) Na_2HPO_4, 3.7% (w/v) KCl, 2.4% (w/v) KH_2PO_4, 4% (w/v) EDTA, 30% (w/v) Trizma base in ultrapure water. pH has to be adjusted to 7.6 with HCl. Add 25% (w/v) trypsin (Gibco) in this solution preheated at 37°C, under stirring. Filter-sterilize on a 0.22-μm membrane. Store at –20°C.
16. G418: stock solution 1000X, 200 mg/mL in ES medium (Roche). Filter-sterilize on a 0.22-μm membrane. Store at –20°C. The dose used for selection has to be determined for each cell line and G418 batch: the minimal dose necessary to kill 100% of nonresistant cells (200 μg/mL for our ES cell line).
17. Gancyclovir: stock solution 20 m*M*. Used at 0.2 μ*M*. Store at –20°C.
18. Culture plates (Corning): diameter 100 mm (B100), 60 mm (B60), 96- and 24-well plates.
18. Electroporation apparatus: Bio-Rad Gene Pulser™ with a capacitance extender.
19. Gamma ray irradiation apparatus.

2.3. Screening of the Resistant Clones

1. PCR lysis buffer (Tween buffer): 50 m*M* KCl, 10 m*M* Tris-HCl, pH 8.3, 1.5 m*M* $MgCl_2$, 0.5% (v/v) Tween-20 in ultrapure water. Sterilize by autoclave. Store at 4°C. Add 0.05% (w/v) proteinase K (PK) extemporaneously.
2. *Taq* DNA polymerase.
3. Polymerase chain reaction (PCR) machine.

2.4. Amplification and Further Characterization of the Positive Clones

2.4.1. Southern Blot

1. Southern blot lysis buffer: 100 m*M* NaCl, 10 m*M* Tris, pH 8.0, 10 m*M* EDTA in ultrapure water. Sterilize by autoclave. Extemporaneously add 0.5% (w/v) sodium dodecyl sulphate (SDS), 0.05% (w/v) PK.
2. Prehybridization (and hybridization) solution: 0.25% (w/v) fat-free milk powder, 4X sodium chloride sodium phosphate EDTA (SSPE [4X SSPE: 600 m*M* NaCl, 40 m*M* NaH_2PO_4, 5 m*M* EDTA]), 1% (w/v) SDS, 0.01% (w/v) denatured salmon sperm in ultrapure water. To be prepared extemporaneously.
3. For hybridization, just add the denatured radioactive probe in the prehybridization solution.

2.4.2. Karyotype

1. Colcemid stock solution 25X (2 μg/mL): demecolcin tested for cell culture (Sigma) diluted in PBS. Store at 4°C.
2. Hypotonic solution: KCl 0.56%. Store at RT.
3. Fixation solution: methanol (3 vol)/ acetic acid (1 vol). To be prepared extemporaneously and kept at RT.
4. Giemsa staining solution: add in this order, 9 mL of water, 1 mL of Giemsa R colorant, and 0.1 mL of Wright. Prepare extemporaneously.
5. Eukitt (O. Kindler GmbH & Co) for slide mounting.

2.4.3. Mycoplasm Detection

1. Standard Mycotect assay from Gibco-BRL.

3. Methods

3.1. Construction of the Homologous Recombination Vector

3.1.1. Structure of the Construction

The vector contains different components (**Fig. 2**):

- A backbone plasmid containing a resistance gene to ampicillin or kanamycin and a replication origin for amplification in bacteria.
- The two arms of homology, which are the DNA genomic sequences surrounding the chromosomal region to be destroyed. The size of the two fragments have to be different: 1–2 kb for the shorter one, 3–6 kb for the longer one. It is easier for the following if one knows the partial or entire sequence of these regions, but this is not absolutely necessary.
- A selection cassette that enables the autonomous expression of a positive selection marker, most of the time a cDNA, providing the resistance to Neomycin (NeoR) under the control of a PGK promoter.

3.1.2. Strategy for the Construction

The different components have to be inserted into the backbone plasmid: the short homology region should be inserted first, in order to avoid the accumulation of restriction sites and to work as long as possible with small plasmids. The different elements in the vector should be ordered as follows, from 5' to 3' (**Fig. 2**): the 5' arm of homology, the positive selection cassette (preferably in the opposite orientation), the 3' arm of homology. A unique restriction site is absolutely required, positioned at either end of the block containing the above elements, for linearization of the construct before electroporation (*see* **Note 2**).

3.1.3. Obtaining the Two Arms of Homology

The simplest way (and the only one developed here) to obtain the fragments is to use PCR amplification on a genomic DNA preparation or on a plasmid containing a fragment of genomic DNA covering the desired region. Cohesive ligations have to be used for all the cloning steps, either taking benefit of some naturally occurring sites or introducing them in the primers used for PCR. When designing the short arm, keep in mind that you have to know a sequence upstream of it, if it is located 5' relative to the region to be deleted (downstream of it, if it is in the 3' position), in order to design a primer for PCR screening of the transfected ES clones.

1. Amplify the two arms by PCR using a reagent able to amplify long DNA fragments with high fidelity (for example Template Long expand, Roche). The

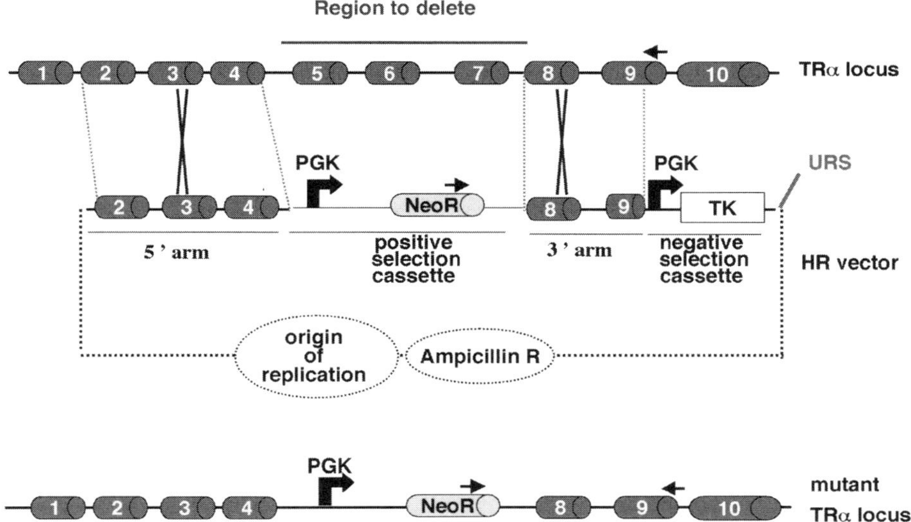

Fig. 2. The homologous recombination vector. The backbone plasmid is figured as a dotted line (.........). For genomic regions, numbered gray cylinders represent the exons, and the thick black line represents the intronic regions. The positive selection cassette contains a cDNA encoding the resistance to G418 (NeoR) (green cylinder) under the control of the PGK promoter (thick arrow). The upper arrows represent the primers used for screening the resistant clones: one in the HR vector, one outside of the short arm. A negative selection cassette can optionally be placed outside, figured as the TK box. A unique restriction site (URS) is placed outside of the arms to linearize the vector before electroporation.

template can either be 10 ng of plasmid or 100 ng of genomic DNA. Elongation time has to be long, approx 1–2 min/kb to be amplified. 25 Cycles are sufficient for plasmid as starting material, 35 cycles are required using genomic DNA preparation.

2. Clone each arm using a TA cloning system and sequence the exonic portions.

3. Insert, one-by-one, the different components by cohesive ligations. The long arm and selection cassette can be cloned into the vector containing the short arm. However, it may be more convenient to sequentially transfer each of the arms into a vector containing the selection cassette. The strategy must be chosen according to the availability of restriction sites. Remember, you have to introduce a unique restriction site at the 3' or 5' position of the recombination block.

4. Digest 40 µg of the final plasmid (the vector for homologous recombination) with the restriction enzyme chosen for linearization for 3 h at 37°C. After checking for a complete cutting, precipitate DNA adding 0.1 vol of sodium acetate 3 *M* and 2.5 vol of ethanol 100%. Wash with ethanol 70% and let the pellet dry in a sterile environment. Resuspend in 40 µL of ultrapure sterile water.

3.2. Homologous Recombination in ES Cells

3.2.1. General Recommendations for ES Cell Culture

ES cells are cultured in complete ES medium on a feeder layer. These feeders are MEFs, which have been irradiated at 45 Gy and seeded in 100 mm culture dishes coated with 0.1% gelatin. Irradiated MEFs cannot be stored more than 1 wk. ES cells have to be trypsinized every 2 d and seeded at a density of 5×10^6 cells per 10 mL. The day after, the medium has to be changed.

3.2.2. Electroporation of ES Cells and Selection of the Resistant Colonies

1. Trypsinize ES cells.
2. Inactivate the trypsin with normal ES medium and spin for 5 min at 800g.
3. Wash the pellet twice with GMEM or OptiMEM.
4. Count the washed cells. Mix 5×10^6 cells with 40 µg of the linearized recombination vector in a total volume of 800 µL GMEM or OptiMEM.
5. Transfer the mixture into a 4-mm electroporation cuvette and perform the electroporation at 260 V and 500 µF.
6. Wait for 20 min before seeding these cells on 5 B100 plates on a MEF layer resistant to the antibiotic used for the positive selection and add 8×10^5 nonelectroporated ES cells per plate.
7. Seed 8×10^5 nonelectroporated ES cells on a MEF layer in a B100 plate as a control.
8. Replace the ES medium 24 h after seeding.
9. Replace the ES medium 14 h later and add the antibiotic used for the selection (200 µg/mL for G418).
10. During the first 3 d of selection, wash the cells with PBS before replacing the medium, in order to discard the maximum of dead cells. During the rest of the selection period, only aspirate the medium and replace it with some fresh medium supplemented with antibiotic everyday.
11. After 4 or 5 d of selection, there should not be any cells left in the control plate, and colonies should appear in the plates seeded with electroporated ES cells.
12. Let the colonies grow up until they occupy the entire field observed using the 100X objective of the microscope, a size usually obtained after 7 to 9 d of selection (*see* **Note 3**).

3.2.3. Isolation and Amplification of the Resistant Colonies

Each colony of resistant ES cells has then to be cloned and amplified (*see* **Note 4**).

1. To pick up the clones, settle the microscope under the laminar flow hood.
2. Aspirate the medium of the B100 plate, wash with PBS, and again add 10 mL of PBS.
3. Each colony has then to be mechanically detached from the plate by scraping around with a tip of a P20 Gilson pipetman. When it is partially detached, just aspirate it in a maximum volume of 15 µL.

4. Mix with 40 μL of TES in an eppendorf tube, dissociate actively by gently pipeting up and down, and wait for 20 min at RT.
5. Each dissociated colony is then individually seeded over into a well of a 96-well plate on a MEF layer full of ES medium.
6. The clones have to be amplified. Just change the ES medium everyday until you estimate that cells are at a normal density in the well (usually 2 to 4 d depending on the initial size of the colony).
7. Trypsinize the cells in the well with 50 μL of TES, inactivate with 100 μL of ES medium, transfer into a well of a 24-well plate on a MEF layer, and fill up the well with ES medium.
8. Change the ES medium every day until you estimate that cells are at a normal density (usually 2 to 3 d).
9. For amplification, trypsinize the cells in the well with 100 μL of TES and add 1 mL of ES medium. 200 μL of this suspension are transferred into a well of a 24-well plate full of ES medium and containing a MEF layer for maintenance. For the screening procedure, the remaining cells are transferred into a well of the same size precoated with 0.1% gelatin, without MEF layer. At this stage, each clone has to be individually identified.
10. One day later, lyse the cells in the screening well (see the protocol below) and replace the medium in the amplification well.
11. The day after, freeze the cells in the amplification well. Trypsinize cells with 100 μL of TES, resuspend them in 400 μL of cold ES medium (4°C), put the plate on ice for 15 min, then slowly add 500 μL of freezing medium, and gently mix. Tightly seal the plate with parafilm and store it at –80°C in polystyrene box for up to 15 d.

3.2.4. Screening of the Resistant ES Cell Clones

3.2.4.1. LYSIS OF THE CELLS

1. Aspirate the medium and wash with PBS.
2. Replace PBS with 200 μL of PCR lysis buffer.
3. Transfer immediately into an Eppendorf and incubate overnight at 56°C under agitation.

*3.2.5. PCR Screening (see **Note 5**)*

1. The lysate (1 μL) is then used to perform the PCR in a total volume of 50 μL.
2. The kind of polymerase used for the reaction and the specific amplification program depend on the size of the fragment to be amplified and should have been set up previously.
3. PCR mixture (15 μL) is then loaded onto an agarose gel.

3.2.6. Amplification and Further Characterization of the Positive Clones

3.2.6.1. AMPLIFICATION

1. Thaw the positives clones as soon as possible after identification.
2. Take the 24-well plate out of the freezer and add 500 μL of prewarmed ES medium (37°C) in the wells containing the positives clones.

3. Move the cell suspension up and down with the pipetman until complete thawing.
4. Place the whole content of the well into a new well on a MEF layer and fill up with ES medium.
5. Replace the medium the next day in order to eliminate the DMSO.
6. Amplify in standard ES cell culture conditions and freeze a few samples in freezing tubes for storage in liquid nitrogen.

3.2.7. Characterization of the Positive Clones by Southern Blot

1. A large amount of ES cell DNA has to be prepared, therefore 2×10^6–10^7 cells should be used as starting material.
2. Seed ES cells from a positive clone on a B100 plate precoated with 0.1% gelatin.
3. When cultures reach high density, wash the plate with PBS, and lyse with 1 mL of Southern blot lysis buffer, into which PK has just been added, transfer into an Eppendorf tube, and incubate overnight at 56°C under stirring.
4. Add 1 mL of phenol-chloroform (v/v) and 100 μL of sodium acetate 3 M, shake, spin for 10 min at 1200g, and transfer the supernatant into a new Eppendorf tube.
5. Add 1 mL of isopropanol, shake and transfer the DNA precipitate into an Eppendorf tube full of ethanol 70%.
6. Transfer the DNA pellet in an empty Eppendorf tube, let it dry, and resuspend it in Tris 5 mM EDTA, 0.1 mM Rnase, 10 μg/mL, pH 7.5, for 1 h at 37°C.
7. Digest 10–15 μg of this genomic DNA with 40 U of the appropriate enzyme in a total volume of 70 μL. Dithiothreitol (DTT) (1 mM) and 1 mM spermidine are added to stabilize some restriction enzymes and to avoid star activity. Incubate at least for 3 h (or overnight) at 37°C.
8. Run the samples (after loading buffer addition) on an agarose gel. Incubate the gel for 15 min in a 0.25 M HCl solution, and then transfer it on a Hybond N$^+$ membrane (Amersham) by capillary transfer under alkaline conditions (0.4 N NaOH) overnight.
9. Wash the membrane twice with a 0.2X SSPE solution and prehybridize it for at least 1 h.
10. Hybridize overnight, with a radiolabeled probe denatured for 5 min at 100°C, wash, and expose. The probe is usually one of the vector arms labeled by random priming–extension (Pharmacia Ready-to-go).

3.2.8. Checking for the Karyotypes

In cell culture, aberrant Karyotypes can arise. Since such abnormality will prevent recombinant ES cells to generate gametes in chimeric animals, it is better to check for the Karyotype of the selected ES clones before they are injected into host blastocysts.

1. Karyotype analysis has to be performed on a subconfluent B60 plate of ES cells cultured on a feeder MEF layer.
2. Replace the medium at least 1 h before beginning the experiment.
3. Add colcemide (0.08 μg/mL) into the medium and incubate from 30 min to 1 h at 37°C.

4. Trypsinize the cells, inhibit the trypsin with ES culture medium, and spin for 5 min at 800g. Resuspend the cell pellet in PBS and spin again. This step has to be repeated twice. The pellet is finally resuspended in 2 mL of hypotonic solution.
5. Let the cells blow up for 10 min at RT and add 2.5 mL of fixative solution.
6. Spin for 5 min at 800g, resuspend the cells in 6 mL of fixative solution at RT, and wash the pellet twice in fixative solution at RT. The pellet is finally resuspended in 0.5 mL of fixative solution.
7. Place the cells for at least 2 h at –20°C.
8. Burst the cells as soon as they are out of the freezer, by letting a few droplets of cell suspension fall onto an tilted slide, pretreated with 70% ethanol.
9. Let the slide dry and stain it with Giemsa for 15 min. Wash and let dry again.
10. Mount the preparation for the observation with Eukitt or aquavitrex and cover with a coverslip.

3.2.9. Checking for the Presence of Mycoplasms

Mycoplasm infection in ES cells may prevent them, after injection into blastocysts, to efficiently colonize the germline. Mycoplasm infection in positive clones can be checked using the standard MycoTect assay from Gibco-BRL. One confluent well from a 24-well plate of ES cells is sufficient to perform the test.

3.3. Conclusion

The efficiency of homologous recombination varies a lot depending on the locus to be targeted and the specific region to be destroyed within this locus. One should also keep in mind that the density of ES cells, when trypsinized for electroporation, and the time when selecting drugs are added, highly influence the ratio of homologous recombinations nonspecific integration, using the same RH vector. For example, the rate of positive clones ranged from 1/300 to 1/5 when we performed the generated the TRα allele.

Thus, so far there is no way to predict the efficiency of homologous recombination. Nevertheless, taking care of some details will help to increase your chances.

4. Notes

1. The Cre-*lox*P system is derived from the P2 bacteriophage. Cre is a recombinase that recognizes some specific sequences, the *lox*P sites, and is able to catalyze the excision of a DNA fragment present between two of these sites arranged as direct repeats (**Fig. 3**). This system is now frequently used to perform tissue- and/or time-specific knock-out.
   Two *lox*P sites are introduced in tandem by homologous recombination in such a way they will flank the region to be deleted, without interfering with gene expression (placed in introns for example). Mice harboring this mutation are then crossed with transgenic mice expressing Cre in a tissue-specific manner. The

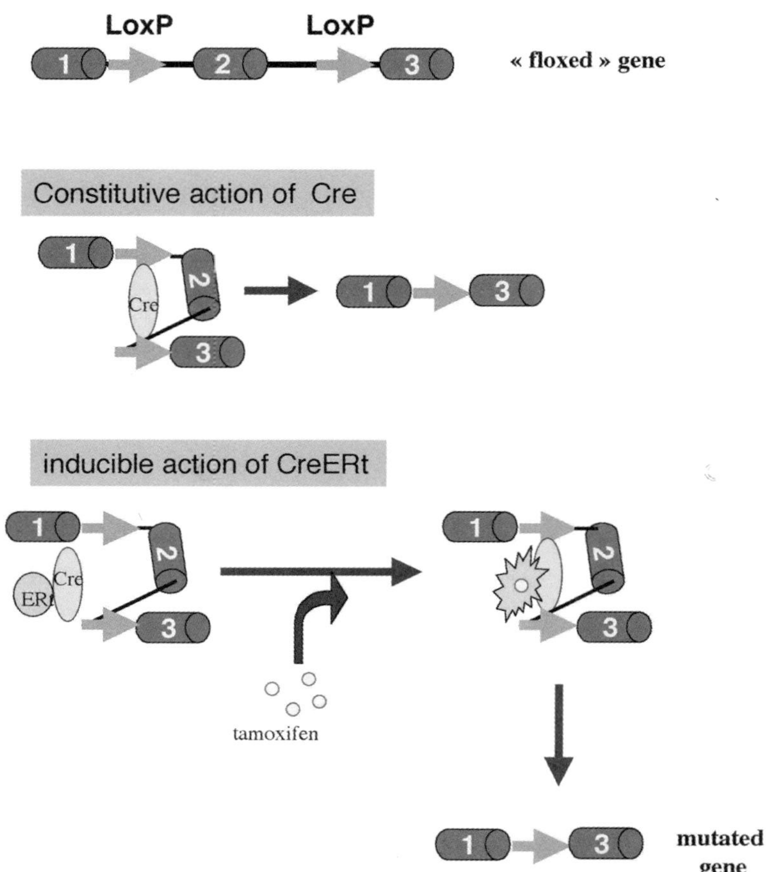

Fig. 3. A technological improvement : the CreLoxP system. LoxP sites are figured as green arrows. The Cre recombinase recognizes two of them arranged as direct repeats and excises the fragment in between. The activity of this enzyme becomes inducible when fused to the ligand-binding domain of the estrogene receptor (ERt) modified to bind only tamoxifen. In the absence of tamoxifen, the enzyme is inactive.

time specificity is obtained using a Cre fused to the estrogen receptor (ER) ligand-binding domain modified to respond only to tamoxifen (CreER^t). In mice expressing this chimeric protein, recombinase activity can be induced by tamoxifen administration *(21)*.

This system can also be very useful to perform some more precise mutations without modifying the structure of the entire locus. It has been particularly helpful in the case of *TRα*, to study the in vivo functions of TRΔα1 and TRΔα2, by preventing their production without altering the expression of neither TRα1 nor TRα2. To do so, a specific deletion of one part of intron 7 containing the pro-

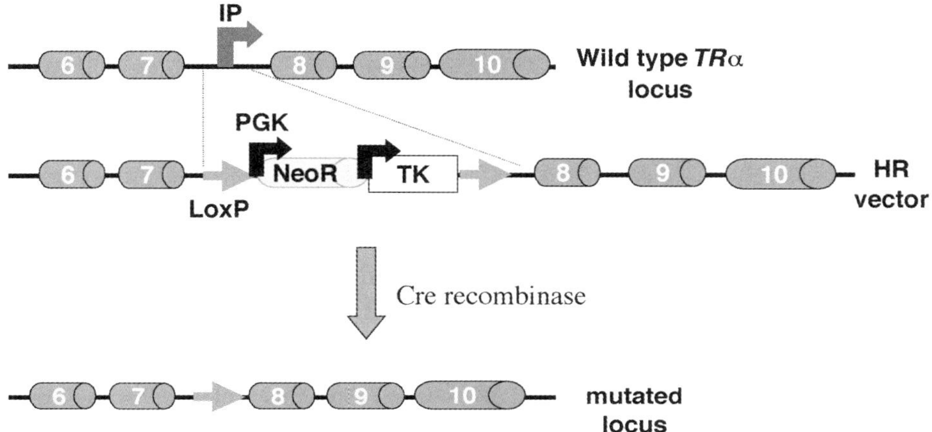

Fig. 4. Homologous recombination for the deletion of the internal promoter (IP) of the *TR*α locus. Signs and symbols are the same as in **Fig. 2**.

moter activity was performed *(22)*. The structure of the RH vector was classical apart from the *lox*P sites flanking the selection cassette (**Fig. 4**). Once positive recombinant ES cell clones were identified (after G418 selection), they were electroporated with 5 µg of a Cre-expressing vector (in the conditions described) to remove the *lox*P=NeoR(TK)=*lox*P cassette, which contained some autonomous promoters and stop signals likely to interfere with the normal transcription and/or translation of the TRα1 and TRα2 products. Gancyclovir was used here as a positive selection, in order to select clones in which TK had been excised. In our hands, excision occurred in 100% of the resistant clones. We have verified that mice homozygous for this mutation, present normal level of TRα1 and TRα2, and strongly reduced level of TRΔα1 and TRΔα2.

This strategy has also been used by Kaneshige et al. *(23)* to introduce a point mutation in the coding region of TRβ, to transform this receptor into a transdominant negative form responsible for the generalized resistance to thyroid hormone syndrome *(24)*. In this case the strategy was quite simple. The easiest way was to clone a very large genomic fragment of the locus (10–15 kb) in a backbone plasmid, to introduce the point mutation (changing 2 or 3 bp) with a recombinant PCR strategy in this genomic sequence, and then to introduce a *lox*=Neo(TK)=*lox* cassette in an intron for recombinant ES cell selection. It is important to place the resistance cassette as close as possible to the point mutation in order to minimize the risk of a recombination event taking place in between. To verify within recombinant ES cells that the point mutation has been retained, it is recommended to introduce a restriction site, either in the mutation or just beside it. Just amplify a DNA fragment surrounding the mutation and cut the PCR product with the appropriate restriction enzyme.

The excision of the selection cassette can either be performed in the ES cells (as described for the deletion of the internal promoter of *TRα*) or in vivo, by crossing mice homozygous for the mutation with transgenic mice expressing the Cre during the very early embryonic stages *(25)* as performed by Kaneshige et al. *(23)*. For all these uses, it is absolutely necessary to sequence all the coding regions present in the RH vector to be sure that the phenotype observed is due to the introduced mutation and not to *Taq* DNA polymerase mistakes.

2. The construction of the homologous recombination vector is a crucial step. The arms of homology in the HR vector have to be as homologous as possible to their cellular genomic counterparts. Use genomic DNA prepared from the same mouse strain as that of the ES cells in which the recombination will be performed. Use a reagent with high fidelity to amplify fragments of genomic DNA. A negative selection cassette can be placed either upstream of the 5' arm or downstream of the 3' arm. This cassette must enable the autonomous expression of a negative selection marker. The most frequently used is the cDNA encoding the herpes simplex thymidine kinase under the control of a PGK promoter. TK will transform the innocuous gancyclovir into a toxic derivative. Gancyclovir is added during the course of the positive selection, but for 5 d only. If integration takes place at a nonspecific site by nonhomologous integration, TK will usually be integrated, resulting in death of the cells upon gancyclovir treatment. In contrast, if homologous recombination occurs, the TK-containing cassette is discarded with the rest of the vector, and the cells are thus insensitive to the addition of gancyclovir in the selection medium. A LacZneo fusion (βgeo) can be used instead of Neo as positive selection gene. The expression of β-galactosidase (β-gal), revealed by 5-bromo-4-chloro-3-indolyl-β-D-galactopyranoside (X-gal) staining can be used to screen the mutant mice. Alternatively, this β geo cassette can be used without promoter, fused to the ATG start site of the targeted cellular gene, present in the 5' arm. With such a construct, the homologously recombined βgeo transcript will be expressed with the same pattern as that of the targeted locus. X-gal staining allows one to follow its expression in ES cells and in the mouse derived from them. However, this strategy is inefficient if the transcription of the targeted gene is too low in ES cells, since it will result in weak resistance of the recombined ES cells to G418 and obviate efficient selection.

3. It is very important to wait until the colony is large enough before cloning them. Look carefully at your plates during the selection. It sometimes happens that colonies stop growing in diameter and grow in thickness. The cells at the periphery then start to differentiate, and the clone will never cover the expected surface. This type of clone should be picked up early. As a general rule, clone only compact colonies, composed of small cells with as few differentiated cells as possible.

4. The number of clones to pick up depends on how many resistant colonies arise and on how many colonies you are able to handle for amplification. Considering that the ratio of positives clones ranges from 0.5–30% (in our hands, when knocking out part of the *TRα* or *TRβ* loci), up to 250 colonies should be screened. In any case, a very large number of resistant clones does not predict high recombi-

nation efficiency, and most of the time this situation occurs when selection has been applied too late.

Interclone contamination can be a problem, especially if you handle numerous clones. Do not use the same pipet to aspirate the medium of the different clones. A convenient trick is to plug a small pipet tip on a Pasteur pipet, and to use a new tip for each well. As soon as you detect a contamination, eliminate this well and treat with bleach.

If you can perform the screening for homologous recombination immediately, you will avoid freezing the wells and, therefore, save time. Indeed the protocol timing allows to obtain the results of the PCR screening before the amplification culture has reached saturation. For this purpose, cell lysis is carried out overnight and PCR the day after. You can discard the negative clones and continue the amplification of the positive clones. You will later check for the correct structure of the recombined locus within the selected clones by PCR or Southern blot analysis.

5. Screening is one of the limiting step that needs particular attention. Many PCRs will be performed at the same time on crude lysates. The PCR should thus be very robust. One good way to settle the parameters of this PCR is to construct a "test plasmid" containing the positive selection cassette together with a genomic DNA fragment corresponding to the short arm extended by a few hundred base pairs. This test plasmid can be diluted (10^{-6}) in wild-type ES cell lysate to provide a control sample for the PCR. PCR conditions (choice of primer pairs, temperature, etc.) can be optimized on this plasmid. However, to be exactly in the right conditions, it is better to electroporate this vector into ES cells, to select and clone one or two resistant colonies, and to optimize the PCR on their DNA extracted in the true conditions. These cells can then be frozen in order to be thawed and cultured during the true screening and serve as a perfect control of every screening step from lysis to PCR. However this test plasmid is a source of PCR contaminations: a complete decontamination (bench and pipets) is required after plasmid amplification.

References

1. Franklyn, J. A. (2000) Metabolic changes in hypothyroidism, in *The Thyroid* (Braverman, L. E. and Utiger, R. D., eds.), Lippincott, Williams, and Wilkins, Philadelphia, pp. 833–836.
2. Legrand, J. (1986) Thyroid hormone effects on growth and development, in *Thyroid Hormone Metabolism*, Dekker, Rotterdam, pp. 503–5334.
3. Scanlon, M. F. and Toft, A. D. (1996) *Regulation of Thyrotropin Secretion* (Braverman, L. E., and Utiger, R. D., eds.) Lippincott-Raven, New York, pp. 220–240.
4. Xu, L., Glass, C. K., and Rosenfeld, M. G. (1999) Coactivtor and corepressor complexes in nuclear receptor function. *Curr. Opin. Gen. Dev.* **9,** 140–147.
5. Forrest, D. and Vennström, B. (2000) Functions of thyroid hormone receptors in mice. *Thyroid* **10,** 41–52
6. Williams, G. R. (2000) Cloning and characterization of two novel thyroid hormone receptor β isoforms. *Mol. Cell. Biol.* **20,** 8329–8342.

7. Abel, E. D., Boers, M.-E., Pazos-Moura, C., et al. (1999) Divergent roles for thyroid hormone receptor β isoforms in the endocrine axis and auditory system. *J. Clin. Invest.* **104,** 291–300.

8. Forrest, D., Hanebuth, E., Smeyne, R. J., et al. (1996a) Recessive resistance to thyroid hormone in mice lacking thyroid hormone receptor beta: evidence for tissue-specific modulation of receptor function. *EMBO J.* **15,** 3006–3015.

9. Fraichard, A., Chassande, O., Plateroti, M., et al. (1997) The *TRα* gene encoding a thyroid hormone receptor is essential for post-natal development and thyroid hormone production. *EMBO J.* **16,** 4412–4420.

10. Gauthier, K., Chassande, O., Plateroti, M., et al. (1999) Different functions for the thyroid hormone receptors TRα and TRβ in the control of thyroid hormone production and post-natal development. *EMBO J.* **18,** 623–631.

11. Göthe, S., Wang, Z., Ng, L., et al. (1999) Mice devoid of all known thyroid hormone receptors are viable but exhibit disorders of pituitary-thyroid axis, growth, and bone maturation. *Genes Dev.* **13,** 1329–1341.

12. Ng, L., Hurley, J. B., Dierks, B., et al. (2001) A thyroid hormone receptor that is required for the development of green cone photoreceptors. *Nature* **27,** 94–98.

13. Wikström, L., Johansson, C., Salto, C., et al. (1998) Abnormal heart rate and body temperature in mice lacking thyroid hormone receptor α1. *EMBO J.* **17,** 455–461.

14. Arpin, C., Pihlgren, M.. Fraichard, A., et al.. (2000) Effects of *T3Rα1* and *T3Rα2* gene deletion on T and B lymphocyte development. *J. Immunol.* **164,** 152–160.

15. Plateroti, M., Chassande, O., Fraichard, A., et al. (1999) Involvement of T3Rα- and β-receptor subtypes in mediation of T3 functions during post-natal murine intestinal development. *Gastroenterology* **116,** 1367–1378.

16. Forrest, D., Erway, L. C., Ng, L., Altschuler, R., and Curran,T. (1996b) Thyroid hormone receptor beta is essential for development of auditory function. *Nat. Genet.* **13,** 354–357.

17. Weiss, R. E., Murata, Y., Cua, K., Hayashi, Y., Seo, H., and Refetoff, S. (1998) Thyroid hormone action on liver, heart and energy expenditure in thyroid hormone receptor β deficient mice. *Endocrinology* **139,** 4945–4952.

18. Macchia, P. E., Takeuchi, Y., Kawai, T., et al. (2000) Increased sensitivity to thyroid hormone in mice with complete deficiency of thyroid hormone receptor α. *Proc. Natl. Acad. Sci. USA* **98,** 349–354.

19. Weiss, R. E. , Forrest, D., Pohlenz, J., Cua, K., Curran, T., and Refetoff, S. (1997) Thyrotropin regulation by thyroid hormone in thyroid hormone receptor beta-deficient mice. *Endocrinology* **138,** 3624–3629

20. Johansson, C., Gothe, S., Forrest, D., Vennstrom, B., and Thoren, P. (1999) Cardiovascular phenotype and temperature control in mice lacking thyroid hormone receptor-beta or both alpha1 and beta. *Am. J. Physiol.* **276,** H2006-H2012.

21. Indra, A. K., Warot, X., Brocard, J., et al. (1999) Temporally-controlled site-specific mutagenesis in the basal layer of the epidermis: comparison of the recombinase activity of the tamoxifen-inducible Cre-ER(T) and Cre-ER(T2) recombinases. *Nucl. Acids Res.* **15,** 4324–4327.

22. Plateroti, M., Gauthier, K., Domon-Dell, C., Freund, J. N., Samarut, J., and Chassande, O. (2001) Functional interference between thyroid hormone receptor α (TRα) and natural truncated TRΔα isoforms in the control of intestine development. *Mol. Cell. Biol.* **21,** 4761–4772.
23. Kaneshige, M., Kaneshige, K., Zhu, X.-G., et al. (2000) Mice with a targeted mutation in the thyroid hormone β receptor gene exhibit impaired growth and resistance to thyroid hormone. *Proc. Natl. Acad. Sci. USA* **21,** 13,209–13,214.
24. Refetoff, S. (1993) The syndromes of resistance to thyroid hormone. *Endocr. Rev.* **14,** 348–399.
25. Lakso, M., Pichel, J. G., Gorman, J. R., et al. (1996) Efficient in vivo manipulation of mouse genomic sequences at the zygote stage. *Proc. Natl. Acad. Sci. USA* **11,** 5860–5865.

3

Transgenic Targeting of a Dominant Negative Corepressor to Liver and Analyses by cDNA Microarray

Xu Feng, Paul Meltzer, and Paul M. Yen

1. Introduction

Thyroid hormone receptors (TRs) and retinoic acid receptors (RARs) are nuclear hormone receptors that play crucial roles in embryogenesis, cell proliferation, differentiation, and metabolism. TRs and RARs repress basal transcription in the absence of ligand and activate transcription upon ligand binding in positively-regulated target genes *(1–3)*. TRs and RARs mediate basal repression through interactions with corepressors, such as nuclear receptor corepressor (NcoR) and silencing mediator for retinoid and thyroid hormone receptors (SMRT) *(4,5)*. NCoR and SMRT are both 270 kDa proteins that have 43% amino acid homology *(4,5)*. These corepressors have two nuclear hormone receptor interaction domains in the carboxy terminus and three transferable repression domains in the amino terminus (RD1, RD2, RD3) *(3,6)*. RD1 and a region downstream of RD3 have been shown to recruit histone deacetylases (HDAC1 and HDAC2) through direct interaction with Sin3B *(3,6–9)*. The TR/corepressor/sin3/HDAC causes hypoacetylation of local histones, which then results in conformational changes in the nucleosome structure and decreases access of enhancers and components of the basal transcriptional machinery to the promoter region and transcriptional start site *(7–10)*. In contrast to basal repression, histone acetylation plays an important role in transcription as p160 coactivators and associated cofactors are recruited by liganded TR, which then forms complexes that contain histone acetyltransferase activity *(11)*. These complexes may then exchange with other complexes that share components with the RNA pol II transcriptional initiation complex *(12,13)*.

From: *Methods in Molecular Biology, Vol. 202: Thyroid Hormone Receptors: Methods and Protocols*
Edited by: A. Baniahmad © Humana Press Inc., Totowa, NJ

Hypoacetylation of histone N-terminal tails has also been associated with transcriptional silencing of chromosome domains and basal repression by many classes of regulators *(10,14–16)*. The NCoR/Sin3/HDAC complex not only mediates transcriptional repression by TR and RAR, but also by other transcription factors, such as Rev-Erb, COUP-TF, MyoD, as well as Mad/Max and Mad/Mxi dimers, *(7–10,17–20)*. Additionally, corepressors can interact with PML-RARα and the PLZF-RARα and LAZ3/BCL6 that are involved in acute promyelocytic leukemia and non-Hodgkin lymphomas, respectively *(21–23)*. Recent evidence also suggests that NCoR and Sin3 may interact directly with the key components of the basal transcriptional machinery to inhibit basal transcription in an alternative repression pathway *(24)*.

Recently, Rosenfeld and colleagues reported a NCoR gene-deleted mouse, which died *in utero* with defects in erythrocyte, thymocyte, and central nervous system (CNS) development *(14)*. In co-injection studies of mouse embryo fibroblasts, they observed reversal of basal repression by retinoic acid receptor using an artificial direct repeat 5 reporter *(14)*. In order to study the role of NCoR on endogenous gene regulation and its postnatal effects, we generated transgenic mice that overexpressed a dominant negative form of NCoR, NCoRi, in liver. In particular, we characterized the effects of NCoR in basal repression by TR and cell proliferation.

1.1. Liver as Target for Thyroid Hormone

The liver is a major target organ of thyroid hormone. It has been estimated that approx 8% of the hepatic genes are regulated by thyroid hormone in vivo *(1)*. We have used a quantitative fluorescent cDNA microarray to identify hepatic genes regulated by thyroid hormone. We sampled 2225 genes on the cDNA microarray, which represents approx 10% of the expressed genes in liver, assuming that the liver transcriptosome contains 10,000–20,000 genes *(25)*. cDNA microarray hybridization is a powerful tool to study hormone effects on cellular metabolism and gene regulation on a genomic scale, as it enables simultaneous measurement and comparison of the expression levels of thousands of genes *(26,27)*. Recently, cDNA microarrays have been used to study the gene expression due to fibroblast differentiation, oncogenesis, aging and caloric restriction of mouse muscle, cell cycle in yeast, and differentiation in *Drosophila (28,30–34)*. Microarray technology is based on an approach where cDNA clone inserts or known oligonucleotide sequences are robotically printed onto a glass slide or nylon membrane and subsequently hybridized to two different fluorescently-labeled probes. The probes are pools of cDNAs, which are generated after isolating mRNA from cells or tissues in two states that one wishes to compare. After resulting fluorescent intensities are detected by a laser confocal fluorescent microscope, the ratio of intensities is measured and analyzed by image processing software *(26–28,35)* (**Fig. 1**).

Our microarray study showed that 55 genes, 45 of which were not previously known to be thyroid hormone-responsive, were found to be regulated by thyroid hormone. Among them, 14 were positively regulated by thyroid hormone, and surprisingly, 41 were negatively-regulated. The gene expression profile showed that T3 regulated a diverse range of liver genes that affect many different aspects of cellular metabolism and function such as gluconeogenesis, lipogenesis, insulin signaling, adenylate cyclase signaling, cell proliferation, and apoptosis *(36)*.

1.2. NCoRi Animal Model to Study Biological Function of NCoR in Liver

Hollenberg et al. *(37)* have reported that a variant form of NCoR, NCoRi, lacks the repression domains in the amino terminus, but retains the nuclear receptor interaction domains. NCoRi is derived from a 3.1 kb cDNA that was originally isolated from a human placental library. NCoRi protein contains the TR interaction domains and surrounding amino acids (AA 1539–2453), but lacks the repressor domains (AA 1–1120) present in full-length murine NCoR. It also has been shown that NCoRi has dominant negative activity on endogenous NCoR as it blocked basal repression by TR in in vitro transfection assays *(37)*. To study the biological function of NCoR in vivo, we constructed an expression vector containing the NCoRi cDNA and the mouse albumin promoter to target expression of NCoRi to the liver *(38)* (**Fig. 2**). NCoRi mRNA was expressed 17 to 146 times higher than endogenous NCoR mRNA in transgenic mouse lines in the euthyroid state *(38)*. Northern blot analysis of various tissues showed that NCoRi mRNA was expressed only in the livers of transgenic mice. Even though the transgenic mice had normal liver weight, appearance, and minimal changes in enzyme activity, the basal transcription of several thyroid hormone-regulated target genes was affected by the overexpression NCoRi in liver. We examined T3-regulated gene expression of hypo- and hyperthyroid transgenic mice. In hypothyroid mice, hepatic expression of Spot 14, bcl-3, glucose 6-phosphatase, and 5' deiodinase mRNA was higher in transgenic mice than littermate controls whereas these genes were induced to similar levels in T3-treated mice (**Fig. 3A**) *(38)*. Derepression was not observed for malic enzyme mRNA expression in hypothyroid mice *(38)*. Thus, NCoRi selectively blocked basal transcription of several thyroid hormone-responsive genes, but had no effect on ligand-mediated transcription. Additionally, NCoRi did not affect ligand-indeoendent activation of at least one thyroid hormone-responsive gene, malic enzyme, suggesting that specific corepressors may mediate basal repression of target genes. We also found that overexpression of NCoRi partially blocked the ligand-independent activation of a negatively regulated target gene, sialyltransferase, in the absence of

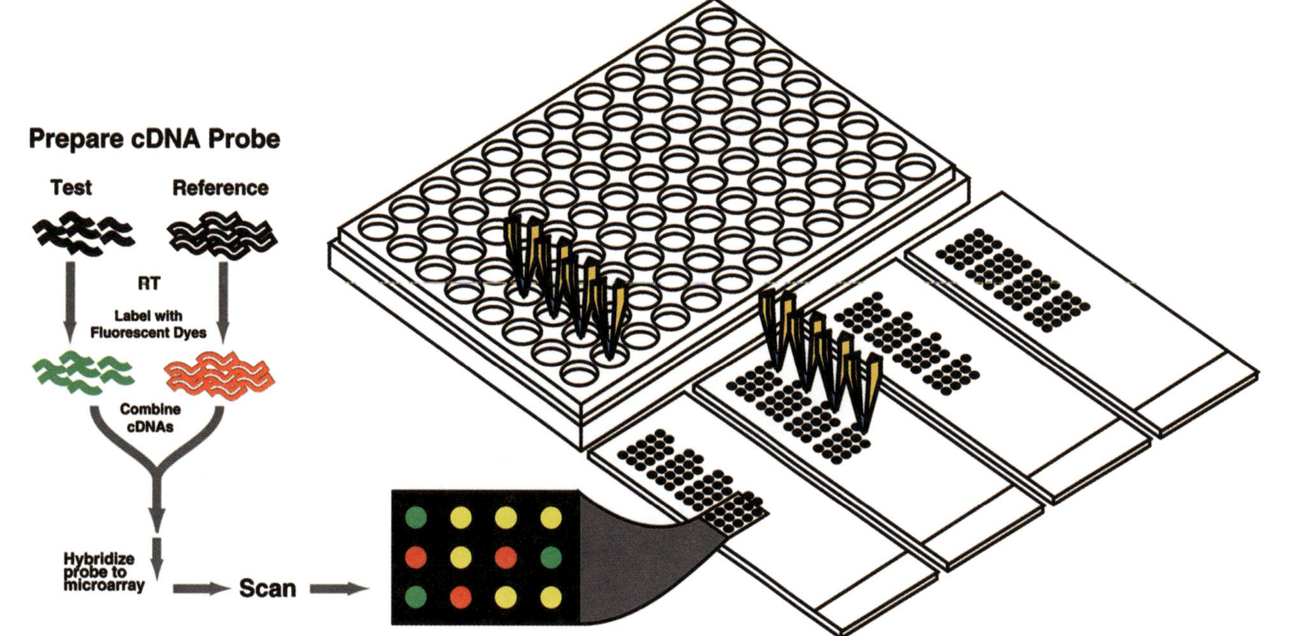

Fig. 1. Schema for cDNA microarray analytes of hepatic RNA from transgenic (test) and wild-type (reference) mice. cDNAs of known identities are spotted as microarrays onto glass slides by pens from robotic arrayer. Hepatic RNA is prepared, purified, and subjected to reverse-transpcription PCR (RT-PCR) as described in **Subheading 3.6.** The cDNAs are labeled with fluorescent dyes (green for test and red for reference), mixed, and then hybridized onto the microarray slide. Fluorescent intensities are then measured and analyzed by a laser confocal microscope. Image analyses was performed using DEARRAY software. Green spots represent genes that are up-regulated in transgenic mice, whereas red spots represent genes that are down-regulated. Yellow spots represent genes that did not vary substantially between transgenic and wild-type samples.

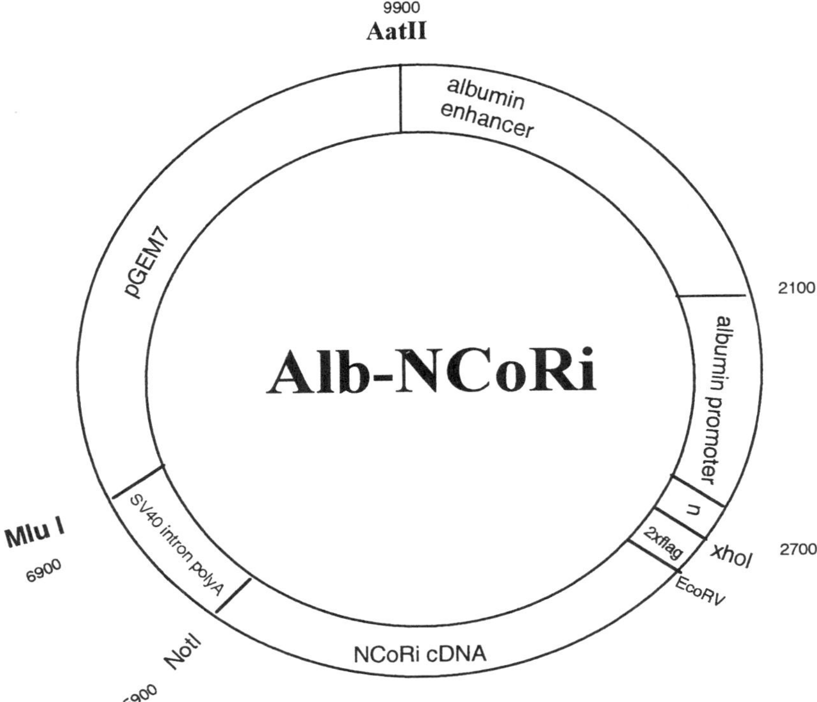

Fig. 2. Schema of vector used to create Alb-NCoRi transgene. NCoRi cDNA (3.1 kb) was inserted into multiple cloning sites of the albumin promoter cassette PGEMLB-SVPA. A FLAG (2X) sequence was also inserted upstream of NCoRi cDNA. The 6.8 kb transgene containing the mouse albumin promoter, NCoRi cDNA, and SV40 intron/poly(A) sequence was generated by digesting the plasmid with *Aat*II and *Mlu*I, and purified as described in **Subheading 3.1.** Abbreviation: µ represents multiple cloning sites.

hormone, which is consistent with the findings of Tagami et al. who showed that NCoR can augment basal transcription of negatively regulated target genes in cotransfection studies *(39)*.

SMRT is another nuclear corepressor expressed in various tissues *(5)*. Like NCoR, SMRT interacts with the unliganded thyroid hormone and retinoic acid receptors via conserved nuclear receptor interaction domains and strongly represses basal transcription in cotransfection studies. We observed that SMRT mRNA was negatively regulated by T3. This suggests that SMRT expression increases in the hypothyroid state when it is involved in basal repression of target genes. Moreover, there was a compensatory increase in endogenous SMRT and NCoR mRNA expression in hypothyroid transgenic mice **(Fig. 4A,B)**. It

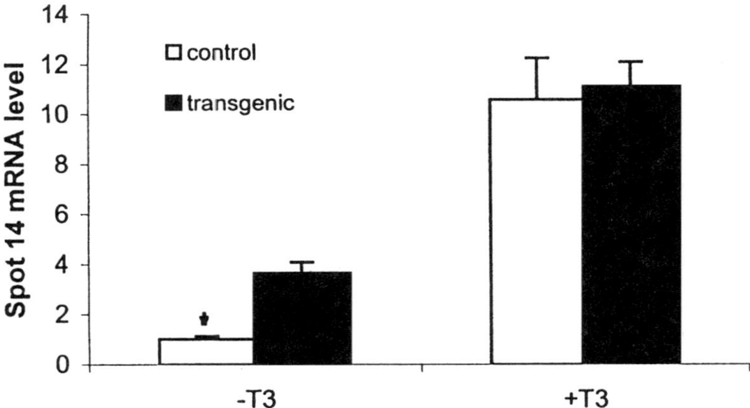

Fig. 3. Spot 14 target gene expression in transgenic and control mice under hypothyroid and hyperthyroid conditions. Poly(A)$^+$ RNA was prepared from livers of hypothyroid control and transgenic mice, and hypothyroid mice treated for 6 h with T3, and Northern blotting performed as described in **Subheading 3.12.** Shown are the means and standard deviation (SD) of 4ENDASH6 samples from control and transgenic mice. All signals were normalized to control 36B4 mRNA expression. Star indicates statistical difference between transgenic and littermate control mice, $P < 0.05$. Similar results were seen for Bcl3, glucose 6-phosphatase, and 5' deiodinase, but not malice enzyme mRNA (*see text*).

currently is not known whether NCoRi directly affects the transcription of these genes or whether derepression of other regulatory genes may alter their expression. Interestingly, a similar compensatory mechanism was observed for coactivators as TIF2 mRNA was increased in the SRC-1 knock-out mouse *(40)*.

To investigate other effects of NCoRi expression on hepatic function, we injected control and transgenic mice intraperitoneally with 5-bromo-2'-deoxyuridine (BrdU). Proliferating cells that incorporated BrdU into their genomic DNA were visualized by staining liver sections with an antibody specific for BrdU **(Fig. 5A)**. As shown in **Fig. 5B**, proliferating cells were increased by 120% in the highest expressing line and 50% in the medium expressing line when compared to control littermates.

cDNA microarray analyses was undertaken to study the patterns of gene expression in the transgenic mouse livers. We observed 28 genes that were up-regulated by more than 2-fold and 10 genes that were down-regulated by more than 60% in two different transgenic lines. We found up-regulation of a number of genes, A fetoprotein, MAP kinase phosphatase-1 (MKP-1), and cyclin A2, which were either markers of proliferation or associated with hepatocyte proliferation *(41–47)* **(Table 1)**. Retinoblatoma (Rb) tumor suppressor gene expression is down-regulated in NCoRi transgenic mouse liver. Rb protein (pRb) blocks the

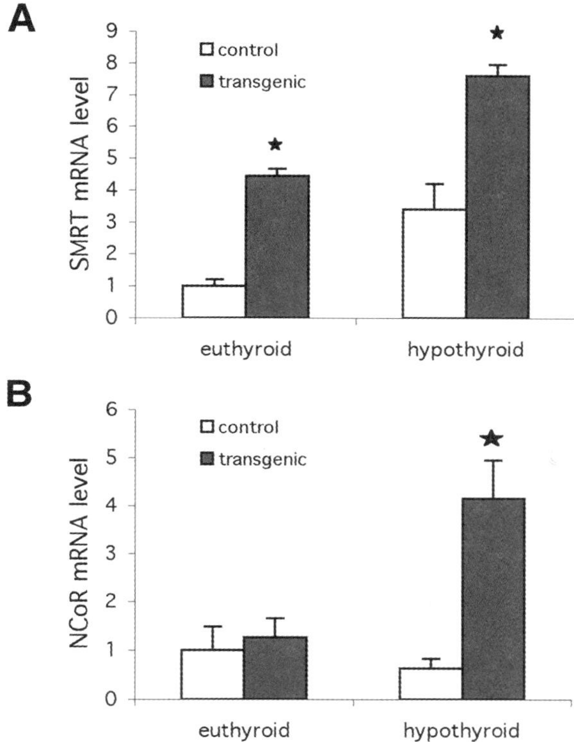

Fig. 4. Endogenous SMRT and NCoR gene expression in livers of transgenic mice. Northern blotting performed as described in **Subheading 3.12.** All signals were normalized to control signal 36B4. Each bar represents the mean of 4 samples. Star indicates statistical difference between transgenic and littermate control mice, $P < 0.05$. (**A**) Bar graph showing hepatic SMRT mRNA levels from hypothyroid and euthyroid transgenic mice and littermate controls. (**B**) Bar graph showing endogenous hepatic NCoR mRNA levels from hypothyroid and euthyroid transgenic mice and littermate controls.

progression from G1 to S phase, and decreased expression or mutations of Rb correlate with cell proliferation and hepatocellular carcinoma *(48,49)*.

1.3. Potential Mechanisms for NCoR Repression of Cell Proliferation

Our NCoRi mouse model showed that NCoR plays an important role in hepatocyte proliferation. Although the precise mechanism by which NCoR represses the expression of these proliferative genes is not known, our findings demonstrate that NCoR has either direct or indirect effects on several key signaling pathways that affect hepatocellular proliferation. There are several cell proliferation pathways that may be affected by NCoR, including the following.

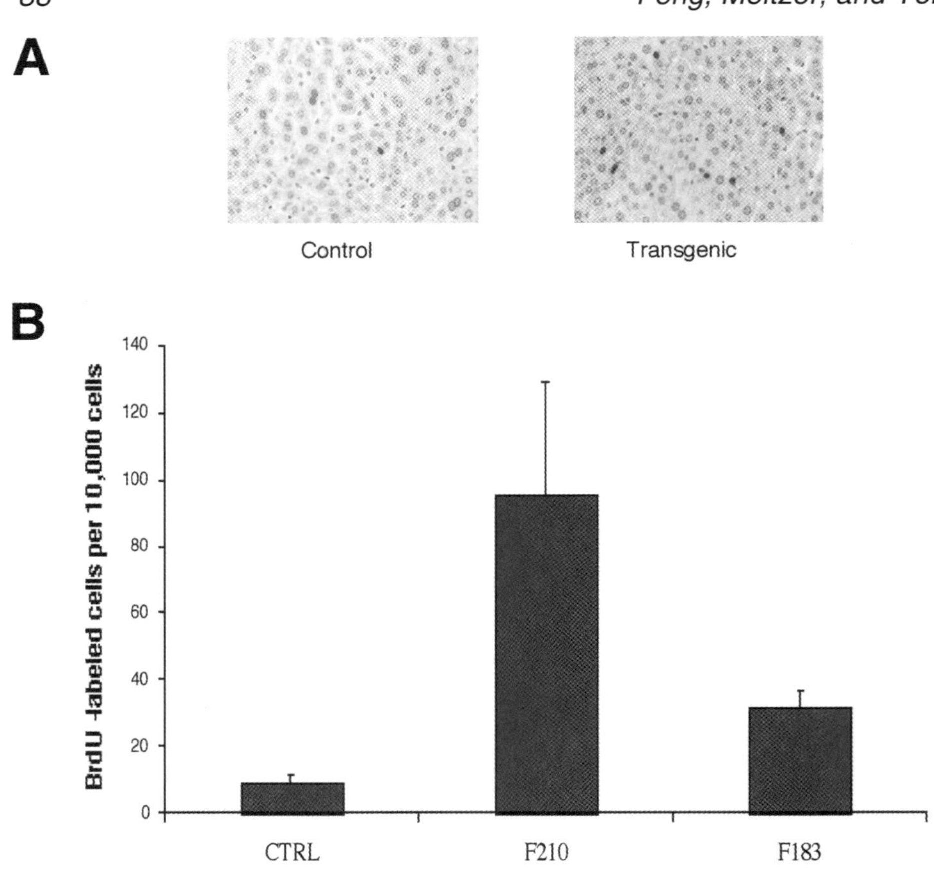

Fig. 5. Effect of NCoRi on hepatic cell proliferation. Control and transgenic mice from lines F210 (high copy number) and F183 (medium-copy number) were injected with BrdU, and hepatic slices were analyzed as described in **Subheadings 2.1., 3.5.,** and **3.13. (A)** Representative micrographs showing BrdU-labeled hepatocytes (red) in liver sections from adult transgenic mice (transgenic) and their littermate controls (control). **(B)** Bar graph showing relative scores of BrdU labeled hepatocytes in 10,000 cells (n = 3 to 4).

1.3.1. Myc-Max-Mad Model

The proto-oncogene c-Myc plays a key role in cell proliferation, differention, and apoptosis. Myc transcripts are rapidly induced upon mitogenic stimulation and are down-regulated during cellular differentiation. These Myc nuclear phosphoproteins promote cell growth and transformation by regulating the transcription of target genes required for proliferation, including ornithine decarboxylase, carbamoylphosphate synthase-aspartate transcarboxylase-dihydroorotase (CAD), and dihydrofolate reductase, cyclin E, p21, cdc25A

Table 1
cDNA Microarray Analysis of Genes Related to Cell Proliferation in Transgenic Mouse Livers

		Fold-Induction	
Clone ID	Gene	F210	F220
387802	α Fetoprotein	3.36	2.48
597868	bcl3	5.05	2.84
515968	cAMP response element modulator	3.57	2.97
552363	Tumor necrosis factor receptor	5.26	4.44
582081	MAP kinase phosphatase-1	4.69	5.56
483145	Insulin-like growth binding protein	3.69	8.4
779163	Cyclin A2	3.68	4.06
463944	Zinc finger protein 90	5.56	3.05
619049	Retinoblastoma-binding protein	0.166	0.157

Hepatic RNA from individual transgenic mice lines F210 (high-copy number) and F220 (medium-copy number) and their littermate controls was prepared, labeled, and analyzed by cDNA microarray analysis as described in **Subheading 3.** and **Fig. 1**. Similar results were obtained in a repeat analysis with different mice from the same lines.

(49–51). Myc promotes cell proliferation, transformation, or apoptosis, requires its dimerization with Max, and binds to its cognate DNA recognization E-box site (CACGTG) *(53)*. It was shown that Mad/Max/Sin3/NCoR/HDAC complexes could repress transcription and potentially inhibit growth *(7–10)*. Thus, Mad/Max and Mxi1/Max dimers can act as antagonists of Myc binding to Myc enhancer sites and repress transcription by recruiting corepressors *(54–55)*. Taken together, these findings suggest that overexpression of NCoRi could disrupt the repression of myc-mediated proliferation as well as other pathways repressed by NCoR.

1.3.2. E2F/Rb Model

The Rb/E2F pathway is parallel pathway to that of Myc and also is involved in cell proliferation *(56)*. The retinoblastoma (Rb) tumor suppressor gene product, pRb, regulates transcriptional events important for cell proliferation. A major target of pRb is the E2F family of transcription factors that control expression of many genes required for DNA synthesis and cell cycle progression. Binding of pRB to E2F species inhibits expression of E2F-regulated genes, resulting in suppression of cell proliferation *(57)*. E2F factors regulate the expression of many genes that encode proteins involved in cell cycle progression and DNA synthesis, including cyclins E and A, cdc2 (cdk1), B-myb, dihydrofolate reductase, thymidine kinase, and DNA polymerase α. Binding of pRb to E2F inhibits E2F's transcriptional activation capacity and, in at least some cases, converts E2F factors from transcriptional activators to transcriptional repressors. Phosphorylation of Rb by D-type cyclin kinases results in the dissociation of Rb from E2F and the expression of the above mentioned E2F-regulated genes *(57)*. The finding that Rb can recruit HDAC1, mSin3 to E2F *(58–60)* also raises the possibility that the Rb/E2F complex may act as a corepressor by hypoacetylation of histones or other target proteins. Down-regulation of expression of Rb in NCoRi transgenic mouse suggests that NCoR might be involved in this process.

Another potential contributor to hepatocyte proliferation is bcl3 *(61)*, which had increased mRNA expression in the transgenic mice. Bcl3, an IκB protein, interacts directly with NFκB homodimers *(62)*. Bcl3 also interacts with transcription integrators such as SRC-1 and CBP/p300 *(61)* and can function as a coactivator of retinoid X receptor and AP1-mediated transcription *(61,63,64)*. Preliminary analysis of the bcl3 promoter showed that it contains potential myc and thyroid hormone response elements *(65)*, suggesting that bcl3 could be a target gene of Myc/Max/Mad and TR. We recently showed that bcl3 is a target gene of thyroid hormone *(35)*. Thus, the interaction of NCoR with members of both the nuclear hormone receptor and the Myc superfamilies raise the possibility that there may be cooperativity between the two pathways.

1.4. Summary

We have generated transgenic mouse lines that overexpress NCoRi in liver. These mice should serve as a useful model for the roles of NCoR in basal repression by thyroid hormones, cell proliferation, as well as gene regulation by steroid hormones and their antagonists. In conjunction with previous studies, they highlight some of the many functions of NCoR (**Fig. 6**).

2. Materials

2.1. Animals

FVB mice used in this experiments were purchased from Charles River Laboratories, Inc. (Wilmington, MA) and maintained in the animal care facility at National Institutes of Health (NIH) according to NIH animal care guidelines. All animals used in these experiments were 8–12 wk old.

2.2. Plasmids

1. Albumin promoter cassette PGEMLB-SVPA (gift from Dr. Jake Liang, NIDDK, NIH) contains mouse albumin promoter, enhancer, multiple cloning sites, Simian Virus 40 (SV40) poly(A) sequence, intron, sequence from PGEM 7 vector (promega).
2. PKCR2-NcoRi (gift of Dr. A. Hollenberg, Beth Israel Hospital, Boston, MA) contains PKCR2 vector sequence and entire human NCoRi cDNA.

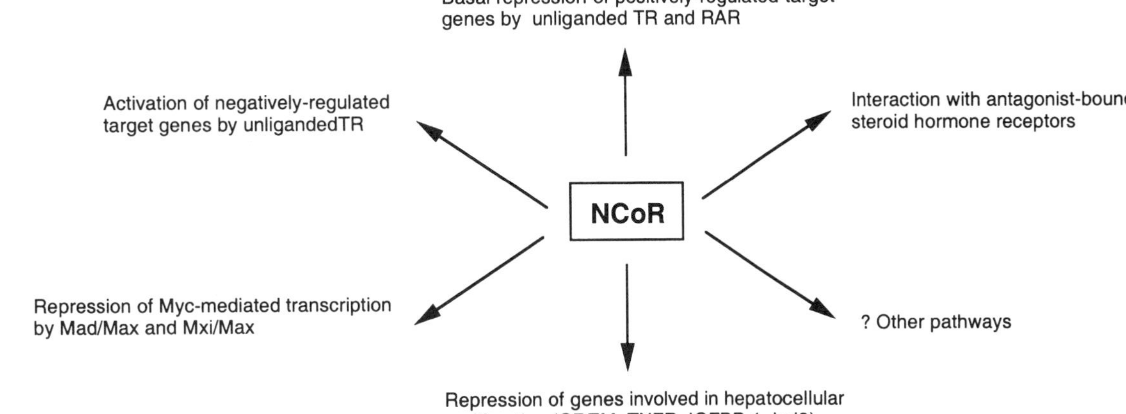

Fig. 6. Model of NCoR action.

3. Alb-NCoRi plasmid was generated by insertion of NCoRi cDNA into mutiple cloning sites of PGEMLB-SVPA *(39)*.

2.3. RNA Extraction

1. Trizol Reagent (cat. no. 15596-018; Life Technologies, Rockville, MD).
2. RNeasy Maxi Kit (cat. no. 75162; Qiagen, Valencia, CA).
3. Chloroform.
4. 3 *M* Sodium acetate, pH 5.2.
5. dATP, dCTP, dGTP, dTTP, 100 m*M* each, store frozen, –20°C (cat. no. 27-2035-02; Pharmacia, Peapack, NJ).
6. pd(T)12–18 resuspend at 1 mg/mL, and store frozen –20°C (cat. no. 27-7858, Amersham Pharmacia Biotech).
7. Anchored oligo-nucleotide primer (anchored; 5'-TTT TTT TTT TTT TTT TTT TTV N-3') resuspend at 2mg/mL, store frozen –20°C (e.g., cat. no. 3597-006, Genosys).
8. Cy3-dUTP, 1 m*M*, and Cy5-dUTP, 1 m*M*, store –20°C, light sensitive Rnase inhibitor, store –20°C (cat. no. N211A; Promega).
9. SUPERSCRIPT™ II RNase H⁻ Reverse Transcriptase Kit, store –20°C, (cat. no. 18064-014; Life Technologies, Rockville, MD)
10. Cot-1 DNA, 1 mg/mL, store frozen –20°C (cat no. 15279-011; Life Technologies, Rockville, MD).
11. 0.5 *M* EDTA, pH 8.0.
12. 1 *N* NaOH.
13. 1 *M* Tris-HCl, pH 7.5.
14. MicroCon 30 (cat. no. 42412; Amicon).
15. Tissue homogenizer (e.g., Polytron PT1200 with Polytron-Aggregate-Dispergier-und-Mischtechnik 147a Ch6014, cat. no.027-30-520-0; Brinkmann Instruments, Inc., Westbury, NY).
16. 10X low T dNTP mixture:

Component	Vol (µL)	m*M* final (1/10) concentration
dGTP (100 m*M*)	25	0.5
dATP (100 m*M*)	25	0.5
dCTP (100 m*M*)	25	0.5
dTTP (100 m*M*)	10	0.5
Diethyl pyrocarbonate (DEPC) H₂O	415	0.2
Total Volume	500	

2.4. Hybridization

1. Microarray hybridization chamber.
2. 50X Denhardt's blocking solution: 5 g Ficoll (type 400), 5g polyvinylpyrrolidine, 5 g bovine serum albumin fraction V, and water to 500 mL. Filter before use.

3. pd(A)40–60 Resuspend at 8 mg/mL, store frozen –20°C (cat. no. 27-7988; Amersham Pharmacia Biotech.).
4. 20X Sodium saline citrate (SSC): 175.3 g NaCl, 88.2 g sodium citrate in 800 mL, pH to 7.0 with 10 M NaOH, fill up to 1000 mL with water, autoclave.
5. Yeast tRNA.
6. 10% Sodium dodecyl sulfate (SDS).
7. Coverslips.
8. Forceps.
9. Coplin jars.
10. 0.2-mL Thinwall polymerase chain reaction (PCR) tubes.
11. 65°C Water bath.
12. Thermocycler for 0.2-mL thinwall PCR tubes.
13. Microarray scanner.
14. Image analysis software.
15. 0.5X SSC/0.01% SDS washing buffer: add 25 mL 20X SSC to 974 mL DEPC water. Sterile-filter on a 0.5-μm filter device. Add 1 mL 10% SDS, mix well, and store at room temperature.
16. 0.06X SSC washing buffer: add 3 mL 20X SSC to 997 mL DEPC water. Sterile-filter on a 0.5-μm filter device. Store at room temperature.
17. 10 mg/mL Human Cot-1 DNA: add 925 μL 100% ethanol and 75 μL 3 M sodium acetate, pH 5.2, to 500 μL Human Cot-DNA (1 μg/μL). Centrifuge at 14,000g to pellet. Aspirate off supernatant and allow to air-dry for 5 min. Resuspend the pellet in 50 μL DEPC water.
18. Yeast tRNA preparation: resuspend yeast tRNA at 10 mg/mL in DEPC (based on the Supplier's quantitation) in a 1.5 mL polypropylene conical centrifuge tube. Add one-half volume of neutralized phenol and vortex mix. Add one-half volume of chloroform, vortex mix and centrifuge 5 min at 10,000g. Transfer aqueous layer to a new 1.5-mL polypropylene conical centrifuge tube. Add 1 vol of chloroform, and vortex mix. Centrifuge for 5 min at 10,000g. Repeat chloroform extraction. Transfer aqueous layer to a new 1.5-mL polypropylene conical centrifuge tube. Add 0.1 vol of 3 M sodium acetate (pH 5.2). Add 2 vol of ethanol. Centrifuge for 5 min at 10,000g. Aspirate off supernatant. Add 1 vol of 70% ethanol. Centrifuge for 5 min at 10,000g. Aspirate off supernatant. Allow pellet to dry. Resuspend in DEPC water at the original volume. Determine the RNA concentration by spectrometry. Dilute to 4 mg/mL and store frozen at –20°C.

3. Methods

3.1. Construction of Liver-Specific Expression Vectors for NCoRi

1. To construct a vector containing the albumin promoter and NCoRi sequence, a 3.1 kb NCoRi fragment is obtained by *Eco*RI digestion from PKCR2-NCoRi (gift of Dr. A. Hollenberg, Beth Israel Hospital, Boston, MA).
2. Insert into multiple cloning sites in the albumin promoter cassette PGEMLB-SVPA (gift from Dr. Jake Liang, NIDDK, NIH), resulting in PGEMLB-SVPA-NCoRi vector.

3. Insert also a Flag (2×) sequence into N terminus of NCoRi cDNA.
4. Verify all sequences by sequencing.

3.2. Creation of Transgenic Mice

1. Alb-NCoRi (6.8 kb) fragment containing the mouse albumin promoter, NCoRi cDNA, and SV40 intron/poly(A) sequence (**Fig. 1A**) is liberated from the plasmid vector PGEMLB-SVPA-NCoRi by digestion with *Aat*II and *Mlu*I.
2. Separate by electrophoresis on 1% agarose gel.
3. Purify through Qiagen gel extraction kit (Qiagen).
4. Follow by Elutip column (Schleicher and Schuell).
5. Microinject with the NIDDK Transgenic Facility, NIH.

3.3. Identification of Transgenic Mice by PCR Genotyping

Transgenic mice carrying Alb-NCoRi are identified by PCR genotyping of tail DNA using primers specific to albumin promoter and NCoRi cDNA.

1. Carry PCR amplification out for 30 cycles.
2. Use 200–400 ng of mouse tail DNA.
3. Each cycle set at 94°C for 30 s, 60°C for 30 s, 72°C for 3 min.

3.4. Southern Blotting Analysis of Genomic DNA from Transgenic Mice

1. Digest 5 μg genomic DNA from mouse-tails with *Eco*RV restriction enzyme.
2. Separate digestion products by agarose gel electrophoresis.
3. Transfer to nylon transfer membrane (Schleicher and Schuell).
4. cDNA probe preparation, hybridization, and washes are carried out as described for Northern blotting (*see* **Subheading 3.12.**).

3.5. Generation of Hypothyroid, Hyperthyroid, and Euthyroid Mice

1. Hypothyroid mice are generated by feeding them with a low iodine (loI) diet supplemented with 0.15% propylthiouracil (PTU) purchased from Harlan Teklad Co. (Madison, WI) for 4 wk.
2. Serum thyroid-stimulating hormone (TSH) measurements (kindly measured by Dr. Samuel Refetoff, University of Chicago, Chicago, IL) show that mice littermate control and transgenic mice treated with PTU were profoundly hypothyroid.
3. Hyperthyroid mice are generated by injecting hypothyroid mice intraperitoneally with 100 μg L-T3 (Sigma) per 100 g mouse body weight in phosphate-buffered saline (PBS) for 6 h.
4. Sacrifice animal.
5. Harvest liver (*see* **Note 1**).

3.6. RNA Preparation for Microarray

1. Total RNA is isolated from individual mouse livers by TRIzol; reagent (Invitrogen) and further purified by RNeasy kit (Qiagen).

2. Add 100 mg of frozen tissue directly to 4 mL of TRIzol and dissociate by homogenization with a rotating blade tissue homogenizer.
3. Add 2/10 vol of chloroform and shake for 15 s.
4. Let stand for 3 min. Centrifuge at 12,000*g* for 15 min at 4°C.
5. Take off the supernatant and add it to a polypropylene tube, recording the volume of the supernatant.
6. Add 0.53 vol of ethanol to the supernatant slowly, while vortex mixing. This will produce a final ethanol concentration of 35% **(Note 2)**.
7. Add the supernatant to an RNeasy maxi column, which is seated in a 50-mL centrifuge tube.
8. Centrifuge at 3000*g* in a clinical centrifuge with a horizontal rotor at room temperature for 5 min.
9. Pour the flow-through back onto the top of the column and centrifuge again. A significant amount of RNA is not captured by the column matrix in the first pass of the RNA containing solution through the column.
10. Discard the flow-through and add 15 mL of RW1 buffer (from Qiagen Rneasy kit) to the column.
11. Centrifuge at 3000*g* for 5 min.
12. Discard flow-through, then add 10 mL of RPE buffer (from Qiagen Rneasy kit).
13. Centrifuge at 3000*g* for 5 min.
14. Discard flow-through, and add another 10 mL of RPE buffer.
15. Centrifuge at 3000*g* for 10 min.
16. Put the column in a fresh 50 mL tube and add 1 mL of DEPC-treated water from the kit to the column.
17. Let stand for 1 min.
18. Centrifuge at 3000*g* for 5 min.
19. Add another 1 mL of DEPC-treated water to the column.
20. Let stand for another 1 min.
21. Centrifuge at 3000*g* for 10 min.
22. Aliquot out 400-µL portions of the column eluate into 1.5-mL Eppendorf tubes.
23. Add 1/10 vol of 3 *M* sodium acetate, pH 5.2.
24. Add 1 mL of ethanol to each tube.
25. Let stand for 15 min.
26. Centrifuge at 12,000*g* at 4°C for 15 min.
27. Wash pellet twice in 75% EtOH, then store at –80°C
28. Resuspend RNA at approx 1 mg/mL in DEPC water.
29. Concentrate to greater than 7 mg/mL by centrifugation on a MicroCon 30 filter unit, centrifuge at 16,000*g*, checking as necessary to determine the rate of concentration. This step removes many residual small to medium sized molecules that inhibit the reverse transcription reaction in the presence of fluorescently derivatized nucleotides.
30. Determine the concentration of RNA in the concentrated sample. Store at –80°C.

3.7. Sample Labeling for Microarray Hybridization

1. Use an oligonucleotide dT(12–18) primer, anneal the primer to the RNA in the following 17-µL reaction:

Component	Cy5 labeling	Cy3 labelling
Total RNA (>7 mg/mL)	80–100 µg	80–100 µg
dT(12–18) primer (1 µg/µL)	1 µL	1 µL
DEPC H$_2$O	to 17 µL	to 17 µL

2. Heat to add 23 µL of reaction mixture containing either Cy5-dUTP or Cy3-dUTP nucleotides, mix well by pipeting, and use a brief centrifuge spin to concentrate in the bottom of the tube: 65°C for 10 min and cool on ice for 2 min (*see* **Note 3**). Reaction Mixture:
   8 µL 5X First strand buffer,
   4 µL 10X Low T dNTPs mix,
   4 µL 1 m*M* Cy5 or Cy3 dUTP,
   4 µL 0.1 *M* Dithiothreitol (DTT),
   1 µL RNasin (30 U/µL),
   2 µL Superscript II (200 U/2µL)
   to 23 µL total vol

3. Incubate at 42°C for 30 min, then add an additional 2 µL Superscript II. Make sure the enzyme is well mixed in the reaction vol and incubate at 42°C for 30–60 min.

4. Add 5 µL of 0.5 *M* EDTA (*see* **Note 4**).

5. Add 10 µL 1 *N* NaOH and incubate at 65°C for 20 min to hydrolyze residual RNA. Cool to room temperature (*see* **Note 5**).

6. Neutralize by adding 25 µL of 1 *M* Tris-HCl, pH 7.5.

7. Desalt the labeled cDNA by adding the neutralized reaction, 400 µL of TE, pH 7.5, and 20 µg of human Cot-1 DNA to a MicroCon 30 cartridge. Pipet to mix and spin for 10 min at 16,000*g*.

8. Wash again by adding 200 µL TE, pH 7.5, and concentrating to about 20–30 µL (approx 5 min at 16,000*g*).

9. Recover by inverting the concentrator over a clean collection tube and spinning for 2 min at 16,000*g* (*see* **Note 6**).

10. Take a 2–3 µL aliquot of the Cy5-labeled cDNA for analysis, leaving 18–28 µL for hybridization.

11. Run this probe on a 2% agarose gel in Tris-acetate electrophoresis buffer (TAE). Note: For maximal sensitivity when running samples on a gel for fluor analysis, use loading buffer with minimal dye and do not add ethidium bromide to the gel or running buffer.

12. Scan the gel on a Molecular Dynamics Storm fluorescence scanner (setting: red fluorescence, 200 µm resolution, 1000 V on PMT).
    Note: Successful labeling produces a dense smear of probe from 400 bp to >1000 bp, with little pile-up of low molecular weight.

3.8. Hybridize Fluorescent cDNA to Slide

1. Determine the volume of hybridization solution required. The rule of thumb is to use 0.033 µL for each mm^2 of slide surface area covered by the coverslip used to

cover the array. An array covered by a 24 × 50 mm cover slip will require 40 μL of hybridization solution (*see* **Note 7**).

2. For a 40 μL hybridization, pool the Cy3- and Cy5-labeled cDNAs into a single 0.2-mL thinwall PCR tube and adjust the vol to 30 μL by either adding DEPC water, or by removing water in a SpeedVac **(Note 8)**.
3. For a 40-μL hybridization, combine the following components:
High Sample Blocking High Array Blocking (*see* **Note 9**).
    30 μL, Cy5+Cy3 probe
    1 μL, Poly d(A) (8 mg/mL)
    1 μL, Human Cot-1 DNA (10 mg/mL)
    6 μL, 20X SSC
    2 μL, 50X Denhardt's blocking solution to 40 μL, 40 μL, total volume
4. Mix the components well by pipeting, heat at 98°C for 2 min in a PCR cycler, cool quickly to 25°C, and add 0.6 μL of 10% SDS. Centrifuge for 5 min at 14,000*g* (*see* **Note 10**).
5. Apply the labeled cDNA to a 24 × 50 mm glass coverslip and then touch with the inverted microarray (*see* **Note 11**).
6. Place the slide in a microarray hybridization chamber, add 5 μL of 3X SSC in the reservoir, if the chamber provides one, or at the scribed end of the slide, and seal the chamber.
7. Submerge the chamber in a 65°C water bath and allow the slide to hybridize for 16–20 h. (*see* **Note 12**).

3.9. Wash Off Uunbound Fluorescent cDNA

1. Remove the hybridization chamber from the water bath, cool, and carefully dry off. Unseal the chamber and remove the slide (*see* **Note 13**).
2. Place the slide, with the coverslip still affixed, into a Coplin jar filled with 0.5X SSC/0.01% SDS wash buffer.
3. Allow the coverslip to fall from the slide and then remove the coverslip from the jar with a forceps.
4. Allow the slide to wash for 2–5 min.
5. Transfer the slide to a fresh Coplin jar filled with 0.06X SSC. Allow the slide to wash for 2–5 min (*see* **Note 14**).
6. Transfer the slide to a slide rack and centrifuge at low rpm (700–1000) for 3 min in a clinical centrifuge equipped with a horizontal rotor for microtiter plates (*see* **Note 15**).

3.10. cDNA Microarray

The cDNA microarrays containing 4000 elements are derived from murine expressed sequence tag (EST) clones obtained from Research Genetics, Inc. (Huntsville, AL) as previously described *(28)*. PCR products generated from these clones were printed onto glass slides as previously described *(28,35,36)*.

3.11. Image Analysis

Hybridized microarray slides are analyzed by a custom-designed laser confocal microscope as previously described *(27,28,35,6)*. Image analyses can be performed using DEARRAY software *(28,36)*.

3.12. RNA Extraction, Probe Preparation, and Northern Blotting Hybridization

1. Total RNA is prepared from mouse livers using TRIzol reagent (Life Technologies, Inc.) according to the manufacturer's instructions.
2. Isolate Poly(A)$^+$ using PolyA tract mRNA isolation kit (Promega).
3. Separate Poly(A)$^+$ RNA (2 μg) on 1% agarose-formaldehyde gel.
4. Transfer to a nylon transfer membrane (Schleicher & Schuell).
5. Probe the blots with gel purified -[α^{32}P]dCTP-labeled fragments, purified by NICK column (Pharmacia) according to manufacturer's instruction.
6. Quantify mRNA signals using ImageQuant software (Molecular Dynamics) and normalize with corresponding thyroid hormone-insensitive 36B4 signals.
7. Determine fold-induction from transgenic mice signal values divided by control mice signal values within the same experiment.

3.13. Histological Examination and Measurement of Cell Proliferation

1. Take mouse liver sections from mouse and fix in 10% neutral-buffered formalin solution (Sigma).
2. Embed the specimens in paraffin.
3. Section at 5 μm.
4. Stain with hemotoxylin-eosin by American Histolab (Rockville, MD).
5. To detect proliferating hepatocytes, inject mice intraperitoneally with BrdU, 50 mg per kg of body weight (Boeringer-Mannhein).
6. After 1 h, kill the mice immediately, and fix liver biopsies in 10% neutral-buffered formalin solution (Sigma).
7. Carry out immunostaining of BrdU and use TUNEL analysis by Molecular Histology Laboratories (Rockville, MD).

4. Notes

1. Control mice are injected with the same volume of PBS alone for 6 h before sacrifice and liver harvest. Euthyroid mice are generated by feeding the mice with normal diet.
2. The ethanol should be added, drop-by-drop, and allowed to mix completely with the supernatant before more ethanol is added. If a high local concentration of ethanol is produced, the RNA in that vicinity will precipitate.
3. Superscript polymerase is very sensitive to denaturation at air/liquid interfaces, so be very careful to suppress foaming in all handling of this reaction.
4. Be sure to add EDTA to stop reaction before adding NaOH.

5. The purity of the sodium hydroxide solution used in this step is crucial. Slight contamination or long storage in a glass vessel can produce a solution that will degrade the Cy5 dye molecule, turning the solution yellow.

6. Conservative approximations of the required spin times. If control of volume proves difficult, the final concentration can be achieved by evaporating liquid in the speed-vac. Vacuum evaporation, if not to dryness, does not degrade the performance of the labeled cDNA.

7. The volume of the hybridization solution is critical. When too little solution is used, it is difficult to seat the coverslip without introducing air bubbles over some portion of the arrayed ESTs, and then the coverslip will not sit at a uniform distance from the slide. If the coverslip is bowed toward the slide in the center, there will be less labeled cDNA in that area and hybridization will be nonuniform. When too much volume is applied, the coverslip will move easily during handling, leading to misplacement relative to the arrayed ESTs and nonhybridization in some areas of the array.

8. If using a vacuum device to remove water, do not use high heat or heat lamps to accelerate evaporation. The fluorescent dyes could be degraded.

9. When there is diffuse background or a general haze on all of the array elements, more of the nonspecific blocker components can be used, as in the High Array Blocking formulation.

10. The fluorescent-labeled cDNAs have a tendency to form small very fluorescent aggregates that result in bright punctuate background on the array slide. Hard centrifugation will pellet these aggregates, allowing you to avoid introducing them to the array.

11. Applying the hybridization mixture to the array and affixing a coverslip is an operation that requires some dexterity to get the positioning of the coverslip and the exclusion of air bubbles just right. It is helpful to practice this operation with buffer and plain slides before attempting actual samples. The hybridization solution is added to the coverslip first, since some aggregates of fluorescent remain in the solution and will bind to the first surface they touch.

12. There are a wide variety of commercial hybridization chambers. It is worthwhile to prepare a mock hybridization with a blank slide, load it in the chamber, and incubate it to test for leaks or drying of the hybridization fluid, either of which cause severe fluorescent noise on the array.

13. As there may be negative pressure in the chamber after cooling, it is necessary to remove the water from around the seals, so that it is not pulled into the chamber and onto the slide when the seals are loosened.

14. The sequence of washes may need to be adjusted to allow for more aggressive noise removal, depending on the source of the sample RNA. Useful variations are to add a first wash, which is 0.5X SSC/0.1% SDS or to repeat the normal first wash twice.

15. If the slide is simply air-dried, it frequently acquires a fluorescent haze. Centrifuging off the liquids results in a lower fluorescent background. As the rate of

drying can be quite rapid, it is suggested that the slide be placed in the centrifuge immediately upon removal from the Coplin jar.

Acknowledgments

We would like to thank Dr. Samuel Refetoff (University of Chicago, Chicago, IL) for measuring TSH levels in the hypothyroid mice, Dr. Anthony Hollenberg (Beth Israel Hospital, Boston, MA) for providing the PKCR2-NCoRi vector, and Dr. Jake Liang (NIDDK, Bethesda, MD) for providing the albumin promoter in PGEMLB-SVPA.

References

1. Oppenheimer, J. A., Schwartz, H. L., Mariash, C. N., Kinlaw, W. B., Wong, N. C. W., and Freake, H. C. (1987) Advances in our understanding of thyroid hormone action at the cellular level. *Endocr. Rev.* **8,** 288–308.
2. Pibiri, M., Ledda-Columbano, G. M., Cossu, C., et al. 2001 Cyclin D1 is an early target in hepatocyte proliferation induced by thyroid hormone (T3). *FASEB J.* **15,** 1006–1013.
3. Seol, W., Mahon, M. J., Lee, Y. K., and Moore, D. D. (1996) Two receptor interacting domains in the nuclear hormone receptor corepressor RIP13/N-CoR. *Mol. Endocrinol.* **10,** 1646–1655.
4. Horlein, A. J., Naar, A. M., Heinzel, T., et al. (1995) Ligand-independent repression by the thyroid hormone receptor mediated by a nuclear receptor co-repressor. *Nature* **377,** 397–404
5. Chen, J. D. and Evans, R. M. (1995) A transcriptional co-repressor that interacts with nuclear hormone receptors. *Nature* **377,** 454–457.
6. Li, H., Leo, C., Schroen, D. J., and Chen, J. D. (1997) Characterization of receptor interaction and transcriptional repression by the corepressor SMRT. *Mol. Endocrinol.* **11,** 2025–2037.
7. Alland, L., Muhle, R., Hou, H., Jr., et al. (1997) Role for N-CoR and histone deacetylase in Sin3-mediated transcriptional repression. *Nature* **387,** 49–55.
8. Heinzel, T., Lavinsky, R. M., Mullen, T. M., et al. (1997) A complex containing N-CoR, mSin3 and histone deacetylase mediates transcriptional repression. *Nature* **387,** 43–48.
9. Nagy, L., Kao, H. Y., Chakravarti, D., et al. (1997) Nuclear receptor repression mediated by a complex containing SMRT, mSin3A, and histone deacetylase. *Cell* **89,** 373–380.
10. Laherty, C. D., Yang, W. M., Sun, J. M., Davie, J. R., Seto, E., and Eisenman, R. N. (1997) Histone deacetylases associated with the mSin3 corepressor mediate mad transcriptional repression. *Cell* **89,** 349–356.
11. McKenna, N. J., Lanz, R. B., and O'Malley, B. W. (1999) Nuclear receptor coregulators: cellular and molecular biology. *Endocr. Rev.* **20,** 321–344.
12. Rachez, C., Lemon, B. D., Suldan, Z., et al. (1999) Ligand-dependent transcription activation by nuclear receptors requires the DRIP complex. *Nature* **398,** 824–828.

13. Ito, M., Yuan, C. X., Malik, S., et al. (1999) Identity between TRAP and SMCC complexes indicates novel pathways for the function of nuclear receptors and diverse mammalian activators. *Mol. Cell Biol.* **3**, 361–370.
14. Jepsen, K., Hermanson, O., Onami, T. M., et al. (2000) Combinatorial roles of the nuclear receptor corepressor in transcription and development. *Cell* **102**, 753–763.
15. Hassig, C. A. and Schreiber, S. L. (1997) Nuclear histone acetylases and deacetylases and transcriptional regulation: HATs off to HDACs. *Curr. Opin. Chem. Biol.* **1**, 300–308.
16. Struhl, K. (1998) Histone acetylation and transcriptional regulatory mechanisms. *Genes Dev.* **12**, 599–606
17. Zamir, I., Harding, H. P., Atkins, G. B., et al. (1996) A nuclear hormone receptor corepressor mediates transcriptional silencing by receptors with distinct repression domains. *Mol. Cell. Biol.* **16**, 5458–5465.
18. Shibata, H., Nawaz, Z., Tsai, S. Y., O'Malley, B. W., and Tsai, M. J. (1997) Gene silencing by chicken ovalbumin upstream promoter-transcription factor I (COUP-TFI) is mediated by transcriptional corepressors, nuclear receptor-corepressor (N-CoR) and silencing mediator for retinoic acid receptor and thyroid hormone receptor (SMRT). *Mol. Endocrinol.* **11**, 714–724.
19. Bailey, P., Downes, M., Lau, P., et al. (1999) The nuclear receptor corepressor N-CoR regulates differentiation: N-CoR directly interacts with MyoD. *Mol. Endocrinol.* **13**, 1155–1168.
20. Grignani, F., De Matteis, S., Nervi, C., et al. (1998) Fusion proteins of the retinoic acid receptor-alpha recruit histone deacetylase in promyelocytic leukaemia. *Nature* **391**, 815–818.
21. He, L. Z., Guidez, F., Tribioli, C., et al. (1998) Distinct interactions of PML-RARalpha and PLZF-RARalpha with co-repressors determine differential responses to RA in APL. *Nat. Genet.* **18**, 126–135.
22. Dhordain, P., Albagli, O., Lin, R. J., et al. (1997) Corepressor SMRT binds the BTB/POZ repressing domain of the LAZ3/BCL6 oncoprotein. *Proc. Natl. Acad. Sci. USA* **94**, 10,762–10,767.
23. Lazar, M. A. (1999) Nuclear hormone receptors: from molecules to diseases. *J. Investig. Med.* **47**, 364–368.
24. Muscat, G. E., Burke, L. J., and Downes, M. (1998) The corepressor N-CoR and its variants RIP13a and RIP13Delta1 directly interact with the basal transcription factors TFIIB, TAFII32 and TAFII70. *Nucl. Acids Res.* **26**, 2899–2907.
25. Zhang, L., Zhou W., Velculescu V. E., et al. (1997) Gene expression profiles in normal and cancer cells. *Science* **276**, 1268–1272.
26. Schena, M., Shalon, D., Davis, R. W., and Brown, P. O. (1995) Quantitative monitoring of gene expression patterns with a complementary DNA microarray. *Science* **270**, 467–470.
27. Duggan, D. J., Bittner, M., Chen, Y., Meltzer, P., and Trent, J. M. (1999) Expression profiling using cDNA microarrays. *Nat. Genet.* **21**, 10s–14s.
28. Khan J., Bittner M. L., Saal, L. H., et al. (1999) cDNA microarrays detect activation of a myogenic transcription program by the PAX3-FKHR fusion oncogene. *Proc. Natl. Acad. Sci. USA* **96**, 13,264–13,269.

29. Iyer, V. R., Eisen, M. B., Ross, D. T., et al. (1999) The transcriptional program in the response of human fibroblasts to serum. *Science* **283**, 83–87.
30. Wang, K., Gan, L., Jeffery, E., et al. (1999) Monitoring gene expression profile changes in ovarian carcinomas using cDNA microarray. *Gene* **229**, 101–108.
31. DeRisi, J., Penland, L., Brown, P. O., et al. (1996) Use of a cDNA microarray to analyse gene expression patterns in human cancer. *Nat. Genet.* **14**, 457–460.
32. Lee, C. K., Klopp, R. G., Weindruch, R., and Prolla, T. A. (1999) Gene expression profile of aging and its retardation by caloric restriction. *Science* **285**, 1390–1393.
33. Spellman, P. T., Sherlock, G., Zhang, M. Q., et al. (1998) Comprehensive identification of cell cycle-regulated genes of the yeast Saccharomyces cerevisiae by microarray hybridization. *Mol. Biol. Cell.* **9**, 3273–3297.
34. Bryan,t Z., Subrahmanyan, L., Tworoger, M., et al. (1999) Characterization of differentially expressed genes in purified Drosophila follicle cells: toward a general strategy for cell type- specific developmental analysis. *Proc. Natl. Acad. Sci. USA* **96**, 5559–5564.
35. Khan, J., Saal, L. H., Bittner, M. L., Chen, Y., Trent, J. M., and Meltzer, P. S. (1999) Expression profiling in cancer using cDNA microarrays. *Electrophoresis* **20**, 223–229.
36. Feng, X., Jiang, Y., Meltzer, P., and Yen, P. M. (2000) Thyroid hormone regulation of hepatic genes in vivo detected by complementary DNA microarray. *Mol. Endocrinol.* **14**, 947–955.
37. Hollenberg, A. N., Monden, T., Madura, J. P., Lee, K., and Wondisford, F. E. (1996) Function of nuclear co-repressor protein on thyroid hormone response elements is regulated by the receptor A/B domain. *J. Biol. Chem.* **271**, 28,516–28,520.
38. Kawamura, T., Furusaka, A., Koziel, M. J., et al. (1997) Transgenic expression of hepatitis C virus structural proteins in the mouse. *Hepatology* **25**, 1014–1021.
39. Tagami, T., Madison, L. D., Nagaya, T., and Jameson, J. L. (1997) Nuclear receptor corepressors activate rather than suppress basal transcription of genes that are negatively regulated by thyroid hormone. *Mol. Cell. Biol.* **17**, 2642–2648.
40. Xu, J., Qiu, Y., DeMayo, F. J., Tsai, S. Y., Tsai, M. J., and O'Malley, B. W. (1998) Partial hormone resistance in mice with disruption of the *steroid receptor coactivator-1 (SRC-1)* gene. *Science* **279**, 1922–1925.
41. Camps, M., Nichols, A., and Arkinstall, S. (2000) Dual specificity phosphatases: a gene family for control of MAP kinase function. *FASEB J.* **14**, 6–16.
42. Phaneuf, D., Chen, S. J., and Wilson, J. M. (2000) Intravenous injection of an adenovirus encoding hepatocyte growth factor results in liver growth and has a protective effect against apoptosis. *Mol. Med.* **6**, 96–103.
43. Snibson, K. J., Bhathal, P. S., Hardy, C. L., Brandon, M. R., and Adams, T. E. (1999) High, persistent hepatocellular proliferation and apoptosis precede hepatocarcinogenesis in growth hormone transgenic mice. *Liver* **19**, 242–252.
44. Camps, M., Nichols, A., and Arkinstall, S. (2000) Dual specificity phosphatases: a gene family for control of MAP kinase function. *FASEB J.* **14**, 6–16.

45. Hong, S. H. and Privalsky, M. L. (2000) The SMRT corepressor is regulated by a MEK-1 kinase pathway: iInhibition of corepressor function is associated with SMRT phosphorylation and nuclear export. *Mol. Cell. Biol.* **20,** 6612–6625.

46. Payraudeau, V., Sarsat, J. P., Sobczak, J., Brechot, C., and Albaladejo, V. (1998) Cyclin A2 and c-myc mRNA expression in ethinyl estradiol induced liver proliferation. *Mol. Cell. Endocrinol.* **143,** 107–116.

47. Blanchard, J. (2000) Cyclin A2 transcriptional regulation: modulation of cell cycle control at the G1/S transition by peripheral cues. *Biochem. Pharmacol.* **60,** 1179–1184.

48. Weinberg, R. A. (1995) The retinoblastoma protein and cell cycle control *Cell* **81,** 323–330.

49. Tabor, E. (1994) Tumor suppressor genes, growth factor genes, and oncogenes in hepatitis B virus-associated hepatocellular carcinoma. *J. Med. Virol.* **42,** 357–365.

50. Dang CV 1999 *c-Myc target* genes involved in cell growth, apoptosis, and metabolism. *Mol. Cell. Biol.* **19,** 1–11.

51. Coller, H. A., Grandori, C., Tamayo, P., et al. (2000) Expression analysis with oligonucleotide microarrays reveals that MYC regulates genes involved in growth, cell cycle, signaling, and adhesion. *Proc. Natl. Acad. Sci. USA* **97,** 3260–3265.

52. Henriksson, M. and Luscher, B. (1996) Proteins of the Myc network: essential regulators of cell growth and differentiation. *Adv. Cancer Res.* **68,** 109–182.

53. Amati, B., Frank, S. R., Donjerkovic, D., and Taubert, S. (2001) Function of the c-Myc oncoprotein in chromatin remodeling and transcription. *Biochim. Biophys. Acta.* **1471,** M135–M145.

54. Rao, G., Alland, L., Guida, P., et al. (1996) Mouse Sin3A interacts with and can functionally substitute for the amino-terminal repression of the Myc antagonist Mxi1. *Oncogene* **12,** 1165–1172.

55. Ayer, D. E., Lawrence, Q. A., and Eisenman, R. N. (1995) Mad-Max transcriptional repression is mediated by ternary complex formation with mammalian homologs of yeast repressor Sin3. *Cell* **80,** 767–776

56. Santoni-Rugiu, E., Falck, J., Mailand, N., Bartek J, and Lukas, J. (2000) Involvement of Myc activity in a G(1)/S-promoting mechanism parallel to the pRb/E2F pathway. *Mol. Cell. Biol.* **20,** 3497–3509.

57. Johnson, D. G. and Walker, C. L. (1999) Cyclins and cell cycle checkpoints. *Annu. Rev. Pharmacol. Toxicol.* **39,** 295–312.

58. Brehm, A., Miska, E. A., McCance, D. J., Reid JL, Bannister, A. J., and Kouzarides T. (1998) Retinoblastoma protein recruits histone deacetylase to repress transcription. *Nature* **391,** 597–601.

59. Lai, A., Kennedy, B. K., Barbie, D. A., et al. (2001) RBP1 Recruits the mSIN3-Histone Deacetylase Complex to the Pocket of Retinoblastoma Tumor Suppressor Family Proteins Found in Limited Discrete Regions of the Nucleus at Growth Arrest. *Mol. Cell. Biol.* **21,** 2918–2932.

60. Magnaghi-Jaulin, L., Groisman, R., Naguibneva, I., et al. (1998) Retinoblastoma protein represses transcription by recruiting a histone deacetylase. *Nature* **391** 601–605.

61. Na, S. Y., Choi, J. E., Kim, H. J., Jhun, B. H., Lee, Y. C., and Lee, J. W. (1999) Bcl3, an IkappaB protein, stimulates activating protein-1 transactivation and cellular proliferation. *J. Biol. Chem.* **274,** 28,491–28,496.
62. Fujita, T., Nolan, G. P., Liou, H. C., Scott, M. L., and Baltimore, D. (1993) The candidate proto-oncogene bcl-3 encodes a transcriptional coactivator that activates through NF-kappa B p50 homodimers. *Genes Dev.* **7,** 1354–1363.
63. Ohno, H., Takimoto, G., and McKeithan, T. W. (1990) The candidate proto-oncogene bcl-3 is related to genes implicated in cell lineage determination and cell cycle control. *Cell* **60,** 991–997.
64. Na, S. Y., Choi, H. S., Kim, J. W., Na, D. S., and Lee, J. W. (1998) an IkappaB protein, as a novel transcription coactivator of the retinoid X receptor. *J. Biol. Chem.* **273,** 30,933–30,938.
65. Rebollo, A., Dumoutier, L., Renauld, J. C., Zaballos, A., Ayllon, V., and Martinez, A. C. (2000) Bcl-3 expression promotes cell survival following interleukin-4 deprivation and is controlled by AP1 and AP1-like transcription factors. *Mol. Cell. Biol.* **20,** 3407–3416.

4

The Role of Thyroid Hormone Receptors in the Heart

Wolfgang H. Dillmann and Bernd R. Gloss

1. Introduction

Thyroid hormone (T3) is an important signaling molecule for cardiac function. Chronic exposure of the heart to either elevated levels of thyroid hormone (hyperthyroidism) or lower thyroid hormone levels (hypothyroidism) have profound effects on cardiac output. Hyperthyroidism increases the risk of cardiac failure dramatically, and hypothyroidism is associated with a diminished contractile performance of the heart, which is frequently compensated by cardiac hypertrophy. The molecular mechanisms that underlie these complex changes in cardiac performance, which are dependent on thyroid hormone are not yet fully understood. We and others have identified key target genes for thyroid hormone that are expressed in the heart and can account for some of the cardiac phenotypes observed in hyper- and hypothyroidism. Because there are reports that thyroid hormone may have so-called extranuclear effects, and there may also be the possibility of indirect effects of thyroid hormone on the heart, we set out to investigate the effects of a mutant thyroid hormone receptor $\beta 1$ expressed in the heart. The mutated cDNA was originally cloned from a human patient by Usala et al. *(1)* and had been sequenced to reveal a 3 bp deletion at positions 1295–1297, which led to a deletion of a threonine at amino acid position 337. The mutated receptor was characterized to have a dominant negative effect when co-expressed with wild-type receptors, which could be explained by the inability of the mutated receptor to bind hormone, at the same time retaining the ability to bind to DNA. Expression of this receptor in the heart should yield insight into the role of the thyroid hormone receptors in the heart, in terms of gene expression and physiological alterations due to impaired thyroid hormone signaling. Our task was, therefore, to bring the mutated cDNA under the control of a promoter that was able to express sufficiently strong.

From: *Methods in Molecular Biology, Vol. 202: Thyroid Hormone Receptors: Methods and Protocols*
Edited by: A. Baniahmad © Humana Press Inc., Totowa, NJ

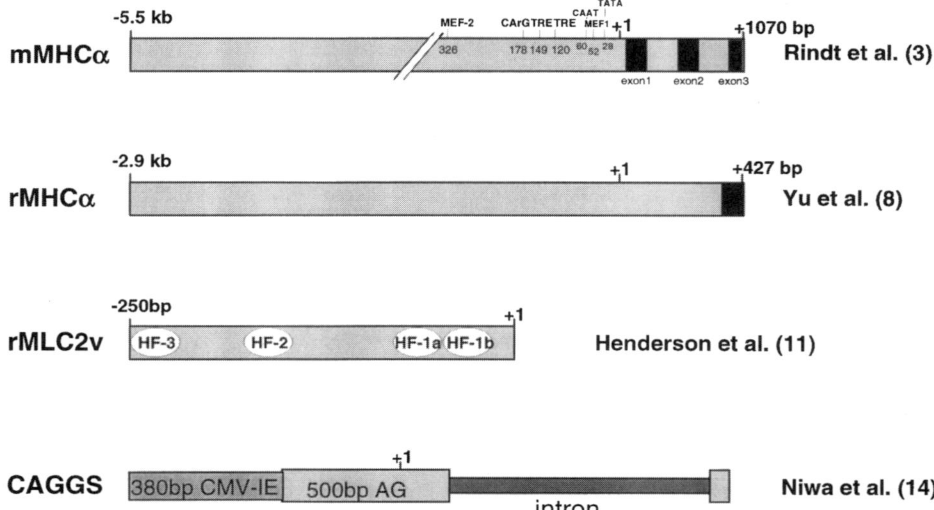

Fig. 1. Schematic illustration of some promoters that are used to drive transgene expression in the heart. The mouse myosin heavy chain α (mMHCα) promoter extends to 5.5 kb upstream of the transcription start indicated by +1. Consensus binding sites for some transcription factors were mapped within the first 326 bp and are indicated with their positions and abbreviations. The noncoding exons 1, 2, and 3 are included in this promoter for efficient transgene expression. The rat myosin heavy chain α (rMHCα) promoter extends to 2.9 kb upstream of the transcription start and includes the noncoding part of its first exon. The rat myosin light chain 2v (rMLC2v) promoter is relatively short and confers ventricular myocyte specificity with only 250 bp upstream of the transcription start. Four binding sites for heart factor (HF) 1a, 1b, 2, and 3 have been located within this promoter. The artificial hybrid promoter CAGGS contains 380 bp of the human CMV immediate early enhancer (CMV-IE) upstream of a chicken ß actin (AG) promoter that contains an intron. MEF-1, muscle enhancer factor 1; MEF-2, muscle enhancer factor 2; TATA, binding region for transcription factor IID; CAAT, binding site for C/EBP factors; TRE, thyroid hormone response element, binding thyroid hormone receptors and the retinoic X receptor; CArG, binding site for the serum response factor (SRF) or CBF-A, factors that are important for sarcomeric actin gene regulation.

This would produce enough mutant thyroid hormone receptor protein in order to impose a dominant negative effect on the endogenous receptor molecules. **Figure 1** gives an overview of some of the promoters that have been successfully used to drive expression of a transgene in the heart of mice. The mouse myosin heavy chain α promoter cloned in Jeffrey Robbins' laboratory has been successfully used to express proteins in the heart *(2–5)*. The rat myosin heavy chain α promoter was originally cloned in Bruce Markham's laboratory and has recently been used to express proteins in mouse hearts *(6–10)*. In a study published by Zhihui Yu, this promoter was used to drive a tetracycline-controlled transactivator (tTA) in transgenic mice. Cross-breeding of such mice with mice that harbor a β-galactosidase transgene under the control of a tetracycline operator revealed substantial β-galactosidase activity throughout the heart, but the response of individual cardiac myocytes was heterogeneous. The mouse myosin light chain promoter was cloned and used in the laboratory of Kenneth Chien to express transgenes in the heart *(11–13)* This promoter is very specific for the myocytes of the ventricles and is not expressed in the atria, making it a useful tool to study effects of a particular gene in the ventricles in the absence of its expression in the atrium. The hybrid promoter that we choose to use was originally described by Niwa et al. *(14)* and is composed of a human cytomegalovirus (CMV) enhancer, which is linked to a chicken β-actin promoter containing an intron, followed by a rabbit β-globin 3' flanking sequence after the cDNA insertion point (*see* **Fig. 2**). This arrangement has proven to yield very high expression in myocytes and also other cell types. We have reported the expression of the mutant thyroid hormone receptor β1 under control of this promoter *(15)* and previously also the *sarcoplasmic-endoplasmic reticulum calcium ATPase (SERCA2)* gene in transgenic mouse hearts *(16)* to demonstrate the beneficial effect of overexpression of this enzyme in various mouse models with impaired cardiac function, in part due to decreased expression of this calcium pump of the sacroplasmic reticulum. **Fig. 3a,b** shows the mRNA and protein levels that were achieved for the SERCA2a expression in transgenic mouse hearts. **Figure 3c** shows a Northern blot with poly(A)$^+$ mRNA from hearts of transgenic mice expressing the mutant thyroid hormone receptor β1, probed for thyroid hormone receptors, and one can see that the expression level of the transgene comes close to the endogenous expression of the thyroid hormone receptors.

2. Materials

1. The vector pCAGGS was originally described by Niwa et al. *(14)* and was intended to be used in a slightly modified form, for transient and stable expression of cDNAs in cultured cells. The promoter and enhancer that were used in this vector were initially tested in front of a *β-galactosidase* gene and were compared with the Rous sarcoma virus (RSV) long terminal repeat (LTR) and the CMV immediate early enhancer in transient transfection experiments into four different cell lines. The unique combination of the chicken β-actin promoter and the CMV immediate early enhancer proved to be a strong driver of linked reporter genes in a variety of cell lines. We have used this vector (*see* **Note 1**) to insert a mutated cDNA of the human thyroid hormone receptor β1 as an *Eco*RI fragment illustrated in **Fig. 3**.
2. The plasmid containing the mutated cDNA of the human thyroid hormone recep-

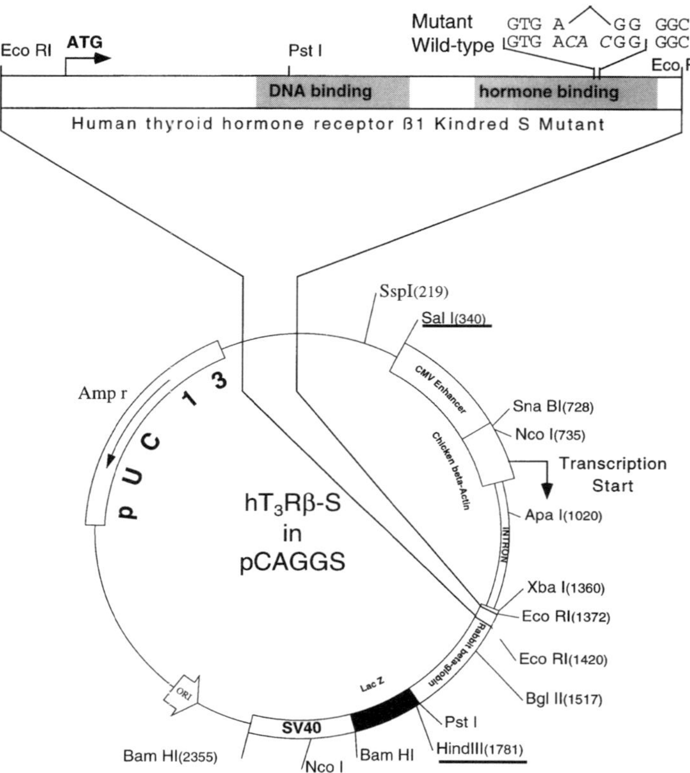

Fig. 2. Map of the plasmid construct from which the transgene, which consists of the CAGGS enhancer–promoter and a mutant human cDNA of the thyroid hormone receptor β1, was derived. The unique *Sal*I and *Hin*dIII sites, with which the plasmid was cut to purify the transgene fragment for oocyte injection, are underlined.

tor β1 was described by Usala et al. *(1)*. The mutated receptor was cloned from reverse-transcribed RNA obtained from a patient with the resistance to thyroid hormone syndrome and determined to belong to the group of the Kindred S mutations. A *Sty*I, *Bgl*II fragment of the mutated, isolated, and cloned cDNA was ligated into the corresponding sites of the plasmid peA 101 described by Weinberger et al. *(17)* to construct a full-length mutated cDNA in the pGEM3 (Promega) vector.

3. Preparation of the plasmid, from which the transgene could be derived, was by CsCl gradient centrifugation according to standard protocols.

4. Purification of the transgene fragment after cleavage with *Sal*I and *Hin*dIII was done with SeaPlaque low melt agarose (*see* **Note 2**). Elutip columns were from Schleicher & Schüll (cat. no. 27370). Elutip High Salt Buffer was at pH 7.3–7.5

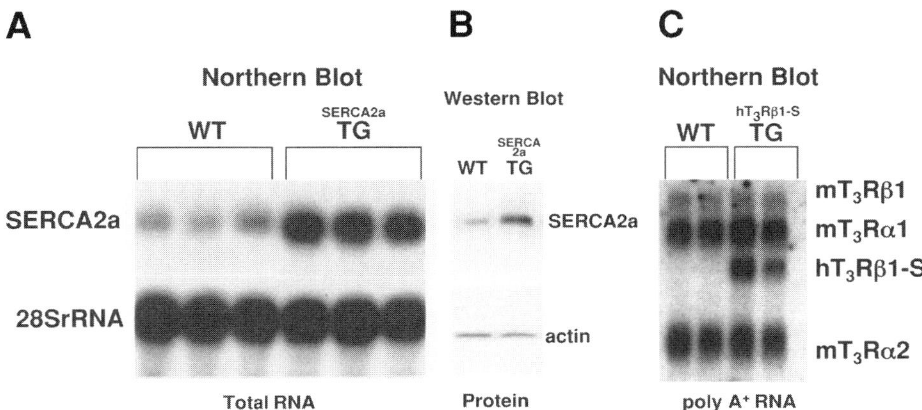

Fig. 3 . Transgene expression in the heart. (**A**) Northern blot with total RNA from the heart of wild-type (WT) and CAGGS driven sarcoplasmic-endoplasmic reticulum calcium ATPase (SERCA2a) transgenic mice (TG). The blot was probed with a rat cDNA for SERCA2a and then, to ensure equal loading, with a probe for 28S-rRNA. (**B**) Homogenates from ventricles of wild-type and SERCA2a transgenic mice were run by sodiumdodecyl sulfate polyacrylamide gel electrophoresis (SDS-PAGE) and processed for Western blot with a SERCA polyclonal antibody and with an antibody against actin as a loading standard. (**C**) Northern blot with polyA+ RNA from wt and CAGGS driven mutant thyroid hormone receptor ß1 transgenic (TG) mice. This blot was probed with a cDNA for the rat thyroid hormone receptor α that recognized both isoforms, including the transcript from the human transgene.

and contained 1 M NaCl, 20 mM Tris-HCl, pH 7.5, 1 mM EDTA. Elutip Low Salt Buffer was at pH 7.3–7.5 and contained 0.2 M NaCl, 20 mM Tris-HCl, pH 7.5, 1 mM EDTA. Injection buffer composition was 7.5 mM Tris-HCl, pH 7.4, 0.15 mM EDTA.

5. Injection of the purified transgene fragment was done with superovulated eggs from C57BL6 X balb/c mice. Injected eggs were transferred to pseudopregnant female Balb/c foster mothers. Mouse husbandry was done in a clean climatized room at 25°C with a 12-h dark/12 h light cycle. Mice were kept in microisolater cages with water and dry food pellets ad libitum.

6. DNA from tail segments was prepared in tail-buffer containing 50 mM Tris-HCl, pH 8.0, 100 mM EDTA, 0.5% sodium dodecyl sulfate (SDS), 0.5 mg/mL proteinase K. Additional reagents for tail DNA preparation are Tris-HCl, pH 8.0, saturated phenol, and phenol–chloroform, ethanol, 3 M Na-acetate, pH 6.5, and 1X TE (10 mM Tris-HCl, pH 8.0, 1 mM EDTA).

7. For the measurement of an electrocardiogram (ECG) in the mouse, the animals were sedated with an intraperitoneal injection of ketamine (6.7 µg/g body weight) and xyalzine (166 µg/g body weight). Four limb restrainers were made by attach-

ing small rubber bands to 18-gauge needles. Conducting leads were made by attaching a thin platinum wire to one end and an adapter-plug on the other end. The analog–digital converter in which the signals from 3 of the 4 leads were first processed was from Beckman Instruments (Model R611) This converter was then linked to a personal computer running a Windaq Software (Data Instruments, Akron, OH).

8. Papillary muscles from mouse hearts were dissected and transferred to a homemade apparatus shown in **Fig. 4**. Oxygenated thyrode solution consists of 136 mM NaCl, 5.4 mM KCl, 1 mM MgCl, 0.33 mM NaH$_2$PO$_4$, 10 mM HEPES, 10 mM glucose, and was at a pH of 7.4. It further contained 30 mM butanedione monoxime and 2.5 mM CaCl$_2$. An isometric force transducer (OPT1L) made by Scientific Instruments (Heidelberg, Germany) was used to measure forces that were generated by the papillary muscle. Electrical stimulation of the papillary muscle was done with a homemade instrument that generated pulses of 5 V, for 0.25 ms duration.

9. Probes for Northern blots were either derived from cDNAs or were synthesized as oligonucleotides. The cDNA for the rat thyroid hormone receptor α was from Howard Towle *(18)*, the SRECA2a cDNA was cloned in our laboratory *(19)*, the human GAPDH cDNA was from American Type Culture Collection (ATCC) No. 78105 (HHCMC32), the oligonucleotide probe for the 28S rRNA was synthesized as an oligonucleotide with the sequence 5'-TGG CAA CAA CAC ATC ATC AGT AGG GT-3', the oligonucleotide probe for mouse myosin heavy chain α was synthesized with the sequence 5'-GAG GCA GGG AAG TGG TGG-3', and the rat myosin heavy chain β oligonucleotide was purchased from Calbiochem (sequence is not disclosed).

10. Total RNA from the heart of the mice was prepared with solution D, containing 4 M guanidinium thiocyanate (Sigma), 25 mM sodium citrate, pH 7.0, and 0.5% sarcosyl. Other reagents that are needed are 2 M sodium acetate, pH 4.0, water-saturated phenol, chloroform, β-Mercaptoethanol, Polytron Homogenizer (Brinkmann Instruments, Westbury, New York, USA).

3. Methods

3.1. Cloning of the Human cDNA

The vector pCAGGS contains two *Eco*RI sites flanking a unique *Xho*I site, located just downstream of the intron of the chicken β-actin promoter (*see* **Note 1**). The vector was opened at the *Eco*RI positions and gel-purified. The pGEM3 vector (Promega), which contained the mutated human cDNA of the thyroid hormone receptor β1 was kindly provided by Stephen J. Usala. The complete coding region could be liberated from the vector backbone by *Eco*RI digestion and was also gel-purified. Following ligation and transformation of bacteria, clones that had the cDNA inserted in the right orientation were isolated, with the 5' end of the cDNA next to the chicken β-actin intron. Positive clones were verified by sequencing and amplified in LB-amp medium to prepare a plasmid

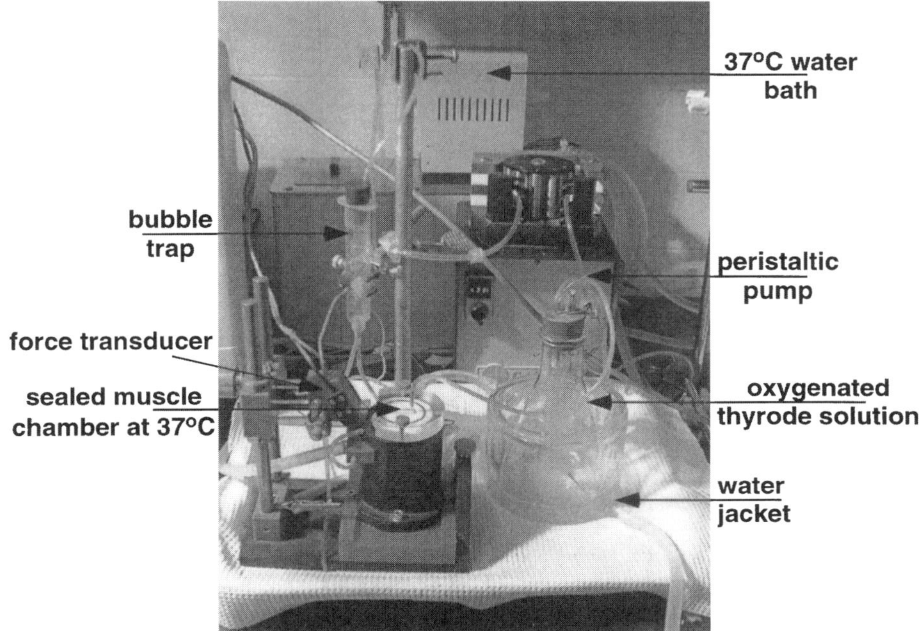

Fig. 4. Device ("Anlage") to measure contractile parameters of left ventricular papillary muscles. One arrow points to a sealed muscle chamber that was perfused with 37°C oxygenated thyrode solution. On the left of this chamber, a force transducer (indicated with an arrow) is attached that converts the mechanical pulling force of the muscle into electrical signals. The instruments to oxygenate, heat, and pump the thyrode solution are depicted. This apparatus was homemade and is a courtesy of Dr. Markus Meyer.

maxiprep followed by a cesium chloride (CsCl) gradient centrifugation to purify the plasmid DNA, according to a standard protocol.

3.2. Transgene Purification

The purity of the DNA fragment to be injected into superovulated eggs is the most critical parameter to the success of the integration of the transgene into the egg cell DNA (*see* **Note 2**). We were, therefore, closely working with our core injection facility to provide a sufficiently clean DNA fragment for injection. This was achieved by following a protocol that was developed by our core transgenic facility. The plasmid DNA can either be isolated using a standard protocol including a CsCl gradient centrifugation or first isolated using a commercial maxiprep-kit followed by a CsCl gradient centrifugation.

1. Digest 100 µg plasmid DNA with the appropriate enzyme(s).
2. Separate the fragment from vector backbone in a 0.8% low melting point agarose gel (Gibco-BRL).
3. Melt the gel piece containing the fragment at 65°C for 10 min. Volume should not exceed 1.5mL.
4. Draw the melted agarose into a syringe filled with 7 mL low salt buffer at 50°C.
5. Attach an Elutip column to the syringe and push out the liquid at about 1 drop/s.
6. Elute the DNA fragment from the Elutip with 400 µL high salt buffer, again at 1 drop/s.
7. Dialyze the eluate in a Slidealyzer (Pierce) against injection buffer overnight.
8. Determine DNA concentration by spectrophotometry. The concentration should be between 5 and 9 µg/mL.
9. Before injection into the oocytes, freeze-thaw the DNA and spin down insoluble material. Use this supernatant to dilute to an appropriate concentration for injection (about 1.8 µg/mL).

3.3. Genotyping of Transgenic Mice

1. After the mice were born from the injected oocytes, they were housed until about 3 wk of age, and then a 1-cm piece of tail from the end was cut and put into an Eppendorf tube. At the same time, these animals were earmarked for later identification.
2. Tail DNA was prepared by incubating the tail at 55°C overnight in 750 µL tail-buffer.
3. The next day, 700 µL Tris-HCl-buffered phenol were added, and the aqueous phase was further extracted with phenol-chloroform.
4. One-tenth of the vol 3 M sodium acetate, pH 6.5, is added, and the DNA is precipitated by adding 700 µL of ethanol.
5. DNA was resuspended in 80 µL of 10 mM Tris-HCl, pH 8.0, 1 mM EDTA.
6. The concentration was determined, and an aliquot of the DNA was used to digest for Southern blot (10 µg) or for polymerase chain reaction (PCR) diagnostic (1 µg).
7. Digestion with *Apa*I and *Pst*I generated a 1579-bp fragment that could be probed on a membrane with a labeled *Pst*I/*Xba*I fragment from the transgene. For PCR, we used a primer pair within the CMV enhancer that yielded a specific amplified product of about 300 nucleotides in length.
8. Animals that carried the transgene were bred to homozygosity, and the copy number of the integrated genes was determined (*see* **Note 3**) by slot-blotting various amounts of transgenic and nontransgenic tail DNA onto a membrane that was subsequently hybridized with a specific thyroid hormone receptor β1 probe consisting of the first 210 nucleotides after the ATG of the human thyroid hormone receptor β1.
9. Using densitometry to analyze the signals from the slot-blot membrane, we calculated that our homozygous founder line, which was subsequently analyzed, had one copy of the transgene integrated into its genome.

3.4. Recording of an ECG in Mice

1. The mouse, being an active small mammal, had to be sedated in order to restrain the animal sufficiently to attach the ECG leads (*see* **Note 4**).
2. Intraperitoneal injection of a ketamine-xyalzine cocktail (50 mg/kg body weight ketamine, 4 mg/kg body weight xylazine) was necessary to attach four limb restrainers made from small rubber bands tied to 18-gauge needles, which were then plugged into a corkboard.
3. The animal was lying on its back with its four limbs at an angle of about 135° relative to the body's longitudinal axis as shown in **Fig. 5**.
4. Four leads, made from a fine stiff platinum alloy, to which a thin electrical wire was soldered, were attached subcutaneously at each limb, close to the trunk. On the other end, the leads were plugged into an analog to digital converter from Beckman (Model R611), which was linked to an IBM-compatible PC running the Windaq software.
5. The ECG was obtained with the leads I, II, III, AVR (A, augmented; V, unipolar; R, right arm), AVL (A, augmented; V, unipolar; L, left arm), AVF (A, augmented; V, unipolar; F, left foot). A representative tracing, which was derived from the data that were recorded, is shown in **Fig. 4** on the computer monitor. Heart rate was calculated from the time point differences between the Q-peaks of an averaged ECG tracing or directly by counting the number of Q-peaks within 1 min. The current peaks were designated on a printed tracing that was obtained from averaging 10 individual tracings. P (duration of atrial depolarization), PQ (duration of excitation progression from the atrium to the ventricle), QRS (duration of depolarization of the ventricles), QT (duration of excitation and repolarization of the ventricles). The way these parameters of the ECG are influenced by hypothyroidism and by expression of a mutant thyroid hormone receptor in the heart is shown in **Table 1**.

3.5. Analysis of Papillary Muscles

1. The papillary muscle in a mouse heart is a very small muscle that is attached to the mitral valve. The advantage of using a papillary muscle for contractile studies is, that the fibers in this muscle are all directed in one direction. It takes some skillful hands to dissect this muscle from the mouse heart under oxygenated thyrode solution and to attach the ends of the muscle preparation to omega-shaped clamps in a home made chamber (courtesy of Dr. Markus Meyer) (*see* **Note 5**).
2. The chamber was perfused with oxygenated thyrode solution at 37°C. The platinum clamps could be moved to stretch the muscle, until it developed maximum force, and were connected to a force transducer, as well as to an impulse generator that stimulated the contraction of the muscle at 2 Hz and 6 Hz in our analyses.
3. The force transducer was either linked to a chart recorder or to a personal computer running the Windaq software, recording force, and time of each contraction, and relaxation at different stimulation frequencies. Impaired thyroid hormone action in the heart, through either hypothyroidism or expression of a

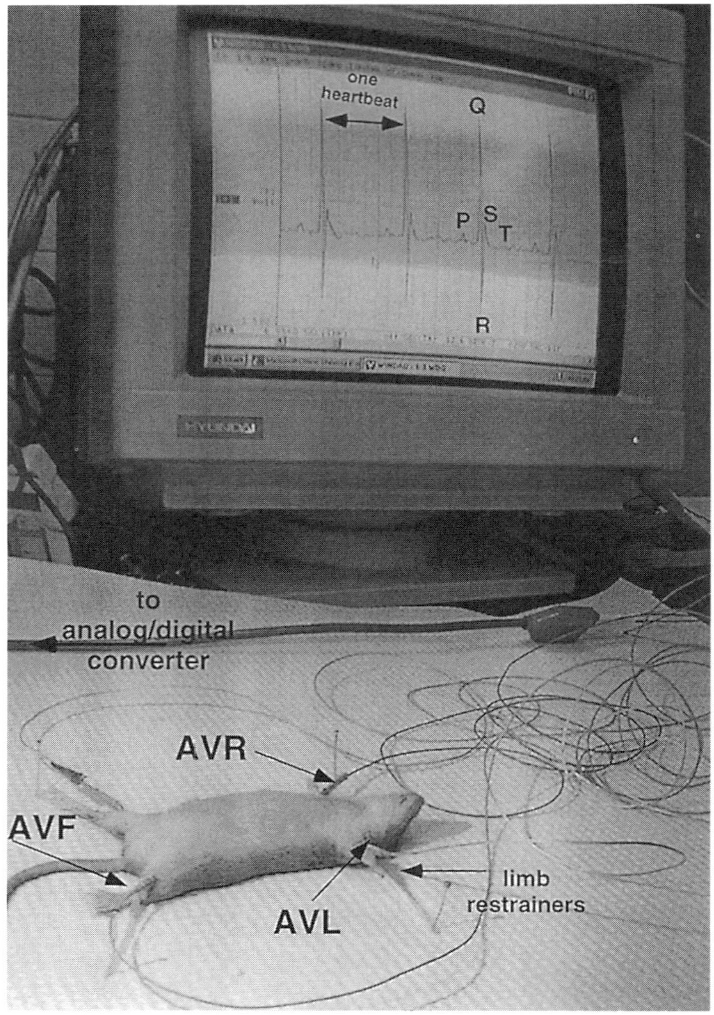

Fig. 5. Recording of an ECG with a sedated mouse in a supine position. The four limb restrainers keep the animal sufficiently still to minimize interference of the recording process due to movements. The designation of the leads is indicated by arrows. Raw, unprocessed tracings of the current changes during the heartbeats are displayed on the monitor. The number of Q-peaks in 1 min represents the heart rate. AVR, AVL, AVF (A, augmented; V, unipolar; R, right arm; L, left arm; F, left foot). P, duration of atrial depolarization; PQ, duration of excitation progression from the atrium to the ventricle; QRS, duration of depolarization of the ventricles; QT, duration of excitation and repolarization of the ventricles.

Table 1
Summary of Values for Heart Rate and ECG Parameters in Wild-Type, Hypothyroid, and Transgenic Mice

Heart rate	WT euthyroid	WT hypothyroid	Transgenic
(beats per min)	301 ± 35	220 ± 12	236 ± 33
P (ms)	15.5 ± 0.7	22.6 ± 1.4	16.0 ± 0.4
PQ (ms)	42.1 ± 1.4	51.3 ± 2.8	44.9 ± 1.0
QRS-duration (ms)	10.6 ± 0.4	12.9 ± 0.5	14.2 ± 0.6
QT (ms)	27.9 ± 1.1	58.5 ± 5.2	33.6 ± 2.0

ECG and heart rate measurements were recorded in sedated mice under supine conditions. WT, wild-type; ECG, electrocardiogram; P, PQ, QRS, and QT, designation of current peaks in the electrocardiogram; ms, milliseconds.

Table 2
Time Course of Contraction of Left Ventricular Papillary Muscles

Group	n	TPT (ms)	Δ	RT50(ms)	Δ
Control	14	42.6 ± 0.9	—	39.8 ± 0.6	—
Hypothyroid	6	64.2 ± 3.1	51%	51.0 ± 2.7	28%
Transgenic	6	51.3 ± 0.5	20.4%	49.4 ± 1.8	24%

Data are means $\pm$ SE. Control, euthyroid wild-type mice; TPT, time to peak tension; RT50, relaxation time to 50% of tension decline; Δ, difference; ms, milliseconds; n, number of animals.

mutant thyroid hormone receptor, leads to alterations of the contractile parameters that can be measured in these papillary muscles and are detailed in **Table 2**. The molecular mechanisms that underlie these thyroid hormone-dependent contractile changes could be due to specific alterations in gene expression in the heart. We, therefore, analyzed specific genes for their expression using Northern blots with total RNA prepared from the heart.

3.6. Preparation of RNA From Mouse Hearts

1. After excision of the heart from the euthanized mouse, the atria were removed (*see* **Note 6**) and the ventricular portion was washed in cold phosphate-buffered saline (PBS), then dried on paper, and then put into an eppendorf tube that was flash-frozen in liquid nitrogen. Hearts can be stored for several months at −70°C.
2. To prepare RNA, the tubes with the hearts were taken from the −70°C freezer, kept on ice, and the tissue was transferred to a 15-mL tube containing 4 mL solution D with 100 mM β-mercaptoethanol.
3. Once thawed, the tissue is then homogenized using a polytron homogenizer until no visible clumps can be seen. To this homogenate the following reagents are added sequentially: 0.4 mL 2 M sodium acetate, pH 4.0, 4 mL water-saturated phenol, 0.8 mL chloroform.

4. The sample was mixed thoroughly and incubated on ice for 15 min, then spun at 4°C at 15,000*g* for 20 min. Four milliliters of the aqueous layer was transferred to a new tube and mixed with 4 mL isopropanol.

5. The RNA is precipitated for 1 h at –20°C and then spun down at 4°C at 15,000*g* for 20 min. The pellet is resuspended in 0.75 mL solution D/100 m*M* β-Mercaptoethanol, transferred to an Eppendorf tube, and mixed with 0.75 mL isopropanol.

6. RNA is again precipitated for 1 h at –20°C and then spun at 4°C in a microcentrifuge for 15 min.

7. The pellet was washed by a 2 min spin with 1mL 70% ethanol (made with diethyl pyrocarbonate [DEPC] treated water) and dried briefly before resuspending it in 0.6 mL 1 m*M* EDTA.

8. Then 0.6 mL phenol–chloroform were added, mixed thoroughly, and spun for 5 min. The aqueous layer (0.4 mL) was transferred into a new tube and 40 μL of 3 *M* sodium acetate, pH 7.0, and 1.1 mL ethanol were added.

9. RNA was precipitated at –20°C for 1 h and spun in a microcentrifuge at 4°C. The pellet was again washed with 70% ethanol by a 2-min spin and then dried, before resuspension, in 40 μL of 1 m*M* EDTA.

10. The concentration of the RNA was determined by spectrophotometry at 260 and 280 nm.

3.7. Probing Northern Blots for Heart-Specific Genes

1. In our analysis of the transgenic mice, we focused on genes that were known to be regulated by thyroid hormone. We used cDNA probes for SERCA2, rat thyroid hormone receptor α and GAPDH. For the myosin heavy chain genes α and β we used oligonucleotides to gain specificity and be able to distinguish the two isoforms (*see* **Note 7**).

2. The hybridization conditions for the labeled oligonucleotides were different from the standard conditions used for the cDNA probes. The hybridization solution was based on a dextran solution containing 10% dextran (Sigma), 1% SDS, 1 *M* NaCl, 50 m*M* Tris-HCl, pH 7.5, and 100 μg/mL sonicated herring sperm DNA.

3. The blots were hybridized in this solution at 65°C overnight, and then washed with 4X standard saline citrate (SSC), 0.05% sodium pyrophosphate at room temperature with 3 to 4 changes of the washing solution, up to a final stringency of 2X SSC, 0.05% sodium pyrophosphate at 50°C.

4. The labeled oligonucleotide for 28SrRNA was hybridized in a different solution containing 1X Denhart's, 0.05% sodium pyrophosphate, 6X SSC and 30 μg/mL herring sperm DNA. The prehybridization solution for this step had five times more Denhart's (×5) and included 0.2% SDS.

5. Agarose gels containing formaldehyde according to a standard protocol for Northern blots were loaded with 15 μg of total RNA from the heart.

6. After electrophoresis, the ribonucleic acids were transferred to a nylon membrane and probed consecutively with different probes.

7. Between the probing, the membranes were washed and stripped of the previous

probe by immersing the filter into water at 90°C for 30 s. Usually the probing for the loading standard with GAPDH or 28SrRNA was peformed last.

4. Notes

1. In general, it is known that the expression of cDNAs in transgenic mice can be problematic, and the expression levels are variable. This is most likely due to different rates of transcription controlled by the promoter driving the cDNA, but also by RNA processing and stability. It seems that the splicing event is benefitial for the correct processing of transcripts and their stability. Therefore, we chose to clone the thyroid hormone receptor cDNA after an intron that is part of the chicken ß actin promoter.

2. We have tried various methods to purify DNA fragments containing transgene expression units for injection into oocytes. The method described here has worked reliably for us and with a variety of transgenes. The excision of the fragment from a highly pure agarose, like the Gibco-BRL brand, is one of the key steps. The use of the Elutip columns avoids precipitaion of the DNA, which is also recommended. Dialysis, using the Slidalyzer system from Pierce, and a freeze-thaw step followed by a spin at approx $12,000g$ purified the DNA sufficiently for oocyte injection and should not be omitted. Despite observing all critical steps, it should be noted that the nature of the transgene could influence the success of creating transgenic animals. In our experience here with the mutant thyroid hormone receptor $\beta1$, we observed a very low frequency of transgenesis in the offspring from injected oocytes. This could point towards a critical role of thyroid hormone signaling during development of the organism in utero. Studies of knock-out animals, in which either thyroid hormone receptor isoforms or both are disrupted, show no grossly abnormal development in utero. Thus, absence of either or both receptors seems to be more tolerable during development than interference of gene transcription by a hormone binding deficient mutant, occupying promoter target sites in a dominant negative fashion.

3. The PCR method with DNA from tails of about 3-wk-old mice worked very reliably, using a primer pair located within the human CMV enhancer that is part of the transgene. We always included a primer pair of the thyroid hormone receptor β to control for the PCR, that gave an amplified product of a different size. PCR positive samples were always confirmed by Southern blot. The determination of the transgene copy number was done by slot-blot. We chose this method because of its higher sensitivity compared with the Southern blot. It should be noted that one has to choose a very specific probe that can detect an endogenous gene, serving as a standard for the signal that a single-copy gene gives, and at the same time can detect the transgene with similar sensitivity. A good range of DNA to blot was between 500 ng and 2 µg. The intensity of the signals recorded on an X-ray film were then quantified by densitometry. Here, one should note that the shortest possible exposure of the film to the membrane usually gives more acurate results.

4. Sometimes it was difficult to obtain consistent recordings from the leads attached to the animal. This could be due to movements of the animal, and, in that case,

sedation of the animal had to be repeated. Another cause for "noise" in the recording were the leads themselves. This could be resolved by shortening the leads (between the animal and the recording unit) and also by using coaxial cables. Alternatively, we were able to record ECG tracings using a portable unit designed for human recordings of ECGs.

5. Some muscle preparations did not yield reliable data, the cause of that is unknown to us. But usually, the preparation of the muscle by one investigator and comparison of 5 to 8 different muscles from control and transgenic groups yielded statistically significant results. It should be noted that the muscle was streched over a period of 30 min to a point where maximal force was developed upon contraction. After this initial stretching period, the muscle could be stimulated to periodically contract 2 or 6×/s. Recording of the different parameters could be taken over a period of 1 h, and one would compare different muscles at the same timepoint after the initial stretching period.

6. The heart is a very heterogeneous tissue, which consists of fibroblasts, endothelial cells, and different kinds of myocytes. In order to be consistent, we removed the left and right atrium from our preparations, because the myocytes of the atria and also their gene expression differ from those of the ventricles. To avoid degradation of the RNA during the thawing process, we immersed the frozen heart in the guanidinium solution at 4°C and homogenized the tissue once it had been sufficiently thawed.

7. Northern blots probed with random labeled cDNA fragments or oligonucleotides labeled by 5' tailing, using terminal transferase and $[\alpha\text{-}^{32}P]dCTP$, will detect relatively abundant messages. Rare messenger RNAs, like those for thyroid hormone receptors and certain ion channels specific for the heart, can only be detected by Northern blot using RNA that was enriched for poly(A)$^+$ mRNAs, or by performing RNase protection assays with species-specific cRNA probes. Note the special hybridization conditions that were necessary to distinguish between the myosin heavy chain α and β isoforms using highly specific oligonucleotides.

Acknowledgments

The authors would like to thank the past and present members of the Dillmann Laboratory who have contributed to these experiments. In particular we are grateful to Drs. Markus Meyer and Wolfgang Bluhm for the setup of the papillary muscle apparatus and the analyses that run with it. We are also thankful to Dr. Susanne Trost for the expert electrophysiology. Finally, the wonderful technical skills of M. Richard Sayen and Eric A. Swanson were vital for most of the experiments.

References

1. Usala, S. J., Menke, J. B., Watson, T. L., et al. (1991) A homozygous deletion in the *c-erbA b thyroid hormone receptor* gene in a patient with generalized thyroid hormone resistance: isolation and characterization of the mutant receptor. *Mol. Endocrinol.* **5,** 327–335.

2. Robbins, J., Palermo, J., and Rindt, H. (1995) In vivo definition of a cardiac specific promoter and its potential utility in remodeling the heart. *Ann. NY Acad. Sci.* **752,** 492–505.

3. Rindt, H., Subramaniam, A., and Robbins, J. (1995) An in vivo analysis of transcriptional elements in the mouse *alpha-myosin* heavy chain gene promoter. *Transgenic Res.* **4,** 397–405.

4. Wickenden, A. D., Lee, P., Sah, R., Huang, Q., and Fishman, G. I., and Backx, P. H. (1999) Targeted expression of a dominant-negative K(v)4.2 K(+) channel subunit in the mouse heart. *Circ. Res.* **85,** 1067–1076.

5. James, J., Zhang, Y., Osinska, H., et al. (2000) Transgenic modeling of a cardiac troponin I mutation linked to familial hypertrophic cardiomyopathy. *Circ. Res.* **87,** 805–811.

6. Passman, R. S. and Fishman, G. I. (1994) Regulated expression of foreign genes in vivo after germline transfer. *J. Clin. Invest.* **94,** 2421–2425.

7. Fishman, G. I. (1995) Conditional trangenics. *Trends Cardiovasc. Med.* **5,** 211–217.

8. Yu, Z., Redfern, C. S., and Fishman, G. I. (1996) Conditional transgene expression in the heart. *Circ. Res.* **79,** 691–697.

9. Redfern, C. H., Coward, P., Degtyarev, M. Y., et al. (1999) Conditional expression and signaling of a specifically designed Gi-coupled receptor in transgenic mice. *Nat. Biotechnol.* **17,** 165–169.

10. Koya, D., Dennis, J. W., Warren, C. E., et al. (1999) Overexpression of core 2 N-acetylglycosaminyltransferase enhances cytokine actions and induces hypertrophic myocardium in transgenic mice. *FASEB J.* **13,** 2329–2337.

11. Henderson, S. A., Spencer, M., Sen, A., Kumar, C., Siddiqui, M. A., and Chien, K. R. (1989) Structure, organization, and expression of the rat cardiac *myosin light chain-2* gene. Identification of a 250-base pair fragment which confers cardiac-specific expression. *J. Biol. Chem.* **264,** 18,142–18,148.

12. Hunter, J. J., Zhu, H., Lee, K. J., Kubalak, S., and Chien, K. R. (1993) Targeting gene expression to specific cardiovascular cell types in transgenic mice. *Hypertension* **22,** 608–617.

13. Hunter, J. J., Tanaka, N., Rockman, H. A., Ross, J., Jr, and Chien, K. R. (1995) Ventricular expression of *amLC-2v-ras fusion* gene induces cardiac hypertrophy and selective diastolic dysfunction in transgenic mice. *J. Biol. Chem.* **270,** 23,173–23,178.

14. Niwa, H., Yamamura, K., and Miyazaki, J. (1991) Efficient selection for high-expression transfectants with a novel eukaryotic vector. *Gene* **108,** 193–199.

15. Gloss, B., Sayen, M. R., Trost, S. U., et al. (1999) Altered cardiac phenotype in transgenic mice carrying the delta337 threonine thyroid hormone receptor β mutant derived from the S family. *Endocrinology* **140,** 897–902.

16. He, H., Giordano F. J., Hilal-Dandan, R., et al. (1997) Overexpression of the *rat sarcoplasmic reticulum* Ca^{2+} *ATPase* gene in the heart of transgenic mice accelerates calcium transients and cardiac relaxation. *J. Clin. Invest.* **100,** 380–389.

17. Weinberger, C., Thompson, C. C., Ong, E. S., Lebo, R., Gruol, D. J., and Evans, R. M. (1986) The *c-erb-A* gene encodes a thyroid hormone receptor. *Nature* **324,** 641–646.

18. Murray, M. B., Zilz, N. D., McCreary, N. L., MacDonald, M. J., and Towle, H. C. (1988) Isolation and characterization of rat cDNA clones for two distinct thyroid hormone receptors. *J. Biol. Chem.* **263,** 12,770–12,777.

19. Rohrer, D. K., Hartong, R., and Dillmann, W. H. (1991) Influence of thyroid hormone and retinoic acid on slow sarcoplasmic reticulum Ca^{2+} ATPase and myosin heavy chain A gene expression in cardiac myocytes. Delineation of cis-active DNA elements that confer responsiveness to thyroid hormone but not to retinoic acid. *J. Biol. Chem.* **266,** 8638–8646.

5

Analysis of Thyroid Hormone-Dependent Genes in the Brain by *In Situ* Hybridization

Juan Bernal and Ana Guadaño-Ferraz

1. Introduction

1.1. Thyroid Hormone and Brain Development

Among the most dramatic actions of thyroid hormone are those exerted on brain development and function. In the adult human brain, a deficiency or excess of thyroid hormone may lead to various psychiatric manifestations, but it is during development when thyroid hormone exerts its most varied and critical actions on neural tissue. In humans, a deficiency of thyroid hormone taking place during a critical period of development may lead to severe mental retardation and also to neurological defects *(1)*. This critical period may extend from the start of the second trimester of pregnancy to the first few months after birth. During this period, the absence of thyroid hormone, if not corrected by early postnatal treatment, leads to irreversible damage with mental retardation. While in utero, the fetal brain is protected from thyroid deficiency by the maternal hormone. Severe thyroid hormone deficiency in the pregnant woman, especially if combined with fetal deficiency, leads to severe neurological deficits in the child that are irreversible even with early postnatal treatment.

In the most studied model of thyroid hormone deficiency, the severely hypothyroid rat, there are no gross alterations of brain morphology. However, there are defects of myelination, alterations of cell migration in the cerebral cortex and the cerebellum, and abnormal differentiation of many neurons, including cholinergic cells and cerebellar Purkinje cells, with severe functional consequences *(2)*.

From: *Methods in Molecular Biology, Vol. 202: Thyroid Hormone Receptors: Methods and Protocols*
Edited by: A. Baniahmad © Humana Press Inc., Totowa, NJ

Elucidation of the molecular basis of thyroid hormone action in the brain is important because it may help to understand how epigenetic factors modify the basic layout of the central nervous system (CNS) and is a necessary requisite for possible therapeutic interventions. Specific nuclear T3 binding can be detected in the rodent brain from about embryonic d 14 (E14) *(3)* and the receptor mRNAs from as early as E11.5 *(4)*. In the human fetus, the receptor is present from at least the onset of the second trimester *(5)*. The role of thyroid hormone in brain development appears to be the fine tuning of gene expression, and several genes have been found to be either up-regulated or down-regulated during particular time windows. In the rat, this sensitive period extends from prenatal d 18 to postnatal d 25–30. In this chapter, we will provide a short summary of the genes that have been identified so far as dependent of the thyroidal status. In some of these genes, thyroid hormone responsive elements (TRE) have been found, indicating that they probably are direct targets of T3 (for instance, MBP, RC3, PCP2, and prostaglandin D2 synthase). In others, no TRE have yet been identified, although in some instances the regulation of gene expression by T3 in cultured cells, in addition to in vivo in the whole animal, also suggests a direct action of the hormone. It is likely that many of the thyroid hormone-dependent genes are regulated indirectly, as a consequence of a distant primary action of T3. Detailed examination of gene expression by thyroid hormone in the brain is out of the scope of the present chapter. What follows is a brief description of some of the gene categories found to be under thyroid hormone regulation (for a review, *see* **ref. 6**).

1.1. Genes of Myelination

Myelination in the CNS is carried out by oligodendrocytes, a special type of glial cells whose terminal differentiation is greatly influenced by thyroid hormone *(7)*. Accordingly all the genes encoding proteins of myelin *(8)*, such as myelin basic protein (MBP), myelin-associated glycoprotein (MAG), proteolipid protein (PLP), and cyclic nucleotide phosphohydrolase (CNP) are under thyroid hormone control in vivo with a similar timing. Dependency of these genes from thyroid hormone is transient, so that in the hypothyroid neonatal rat there is a delayed accumulation of mRNA and protein, but eventually they reach normal levels even in the absence of thyroid hormone treatment and become thyroid hormone independent.

1.2. Mitochondrial Genes

Thyroid hormone affects mitochondrial gene expression in vivo and in vitro. In brain mitochondria several nuclear-encoded (cytochrome c oxidase subunits IV and VIc *(9)*, and a mitochondrial protein import receptor, homologue to the fungal Tom70 *(10)* and mitochondrial-encoded (subunit 3 of NADH dehydrogenase *(11)*, subunit III of cytochrome c oxidase, and 12S and 16S rRNAs *(9)* mRNAs have been observed to be down-regulated in hypothyroidism.

1.3. Neurotrophins and their Receptors

NGF, trkA, and p75[NTR] mRNAs are decreased in hypothyroid rats and increased after thyroid hormone treatment *(12)*. These genes are responsive both in neonatal and in adult rats. Other neurotrophins, such as NT-3 are also under thyroid hormone regulation in specific cells of the cerebellum *(13)*.

1.4. Cytoskeletal Components

Thyroid hormone differentially affects the expression of tubulin isoforms. For example, Tα1 tubulin is down-regulated by thyroid hormone, whereas the β4 isoform is up-regulated *(14)*. Microtubule-associated proteins are also affected by hypothyroidism and thyroid hormone treatment. Tau splicing is regulated by thyroid hormone, which promotes the transition from the juvenile to the mature form *(15)*, and the accumulation of MAP-2 protein is delayed in the hypothyroid animal.

1.5. Transcription Factors and Splicing Regulators

Some transcription factors have been found to be under thyroid hormone influence in vivo, including the early response gene *NGFI-A (16)*, the orphan receptor RORα*(17)*, and basic transcription element-binding protein (BTEB) *(18)*. The regulation of transcription factors expression by thyroid hormone obviously should have far reaching physiological consequences. However, it is unknown how the regulated transcription factor target genes are modified in response to thyroid hormone deficiency or excess. In addition to regulation of transcription, recent data suggest that thyroid hormone could be involved in splicing mechanisms by modifying the expression of splicing regulators *(19)*.

1.6. Extracellular Matrix Proteins

Extracellular matrix proteins such as tenascin-C, laminin and reelin, and cell adhesion molecules, such as NCAM and L1 are also regulated by thyroid hormone during the late prenatal and immediate postnatal period in the rat *(20–22)*. Tenascin-C, NCAM, and L1 are down-regulated by the hormone, whereas laminin and reelin are up-regulated. The control of these proteins, especially reelin, is important for proper neuronal migration in the cerebral cortex, cerebellum, and other brain areas.

1.7. Genes Encoding Proteins Involved in Intracellular Signaling

The expression of a number of proteins involved in cell signaling in different parts of the CNS are under thyroid hormone control. These include RC3/

neurogranin, a protein kinase C substrate and calmodulin-binding peptide *(23)*, calmodulin-dependent kinase IV *(24)*, prostaglandin D2 syntase *(25)*, and Rhes, a novel Ras protein homolog preferentially expressed in the striatum *(26)*.

1.2. Methodology Used to Study the Influence of Thyroid Hormone on Regional Brain Gene Eexpression

The study of gene expression in the CNS requires techniques that allow regional and cellular resolution. To examine the regional and cellular distribution of specific mRNAs, *in situ* hybridization should be used. This technique is based on the hybridization of a labeled probe to the mRNA present in the tissue. The probe can be either DNA or RNA. DNA probes can be prepared from cDNA or from oligonucleotides. The highest specificity and signal-to-noise ratios are achieved with RNA probes, or riboprobes. These are complementary RNAs synthesized in vitro with phage RNA polymerases using as template the specific cDNA. Depending on the position of the polymerase promoter relative to the cDNA sequence, sense or antisense riboprobes can be synthesized. The antisense RNA will hybridize to tissue mRNA, whereas the sense RNA will not and is used as a control of the hybridization. There are a number of techniques to label the probes using either isotopic or nonisotopic methods (for descriptions of different methods and applications, *see* **refs.** *27–29*). Among the nonisotopic, digoxygenin or biotinylated probes are the most commonly used. In our experience, the use of radioactive probes, in general, provides more sensitivity and specificity than nonisotopic probes, although the latter allows the visualization of cell morphology. Radioactive probes also allow for quantification of the signal. If a nonisotopic method is used for simplicity, convenience, or to avoid radioactive waste, it is advisable to compare the hybridization pattern with that obtained using an isotopic probe.

In this chapter, we describe an isotopic in situ hybridization method which is currently used in our laboratory. Sections from the rat brain are hybridized in flotation with [^{35}S]UTP-labeled riboprobes. The use of floating sections results in better signal-to-noise ratios than hybridizing the sections previously immobilized on glass slides, although the analysis of fragile and/or small sections, for example those obtained from embryonic tissue, is more difficult. After hybridization, the signal can be detected by autoradiography, using X-ray films, or photographic emulsions. Autoradiography using X-ray films provides a regional pattern of expression of the target. Emulsion autoradiography is used when cellular resolution is desired. If combined with immunohistochemistry, the cells expressing the gene of interest can be identified using specific antibodies.

2. Materials

2.1. Handling the Animals

1. 1 M phosphate buffer (PB), pH 7.2–7.4: Mix 610 mL of 1 M Na_2HPO_4 and 390 mL of 1 M NaH_2PO_4 to final volume of 1 L.
2. Phosphate-buffered saline (PBS): for 1 L of a 10X stock solution dissolve 80 g NaCl, 2 g KCl, 2 g KH_2PO_4, 11.5 g Na_2HPO_4, adjust to pH 7.4, and add water to 1 L.
3. Thyroid hormones: 3,3',5-triiodo-L-thyronine (T3) sodium salt and L-thyroxine (T4) sodium salt. (Sigma, St . Louis, MO).
   Concentrated stock solutions of these hormones can be prepared in advance and kept at –20°C for at least 3 mo: T4 (1 mg/mL) or T3 (4 mg/mL) are dissolved in 50 mM NaOH, with vigorous vortex mixing, and aliquots are stored at –20°C wrapped in aluminum foil. Working solutions are prepared before administration to the animals by diluting the stock solutions in PBS containing 0.1% (w/v) bovine serum albumin (BSA) as follows:
   a. T4: add 20 µL of stock solution to 980 µL PBS-BSA (final concentration 2 µg T4/100 µL).
   b. T3: add 250 µL of stock solution to 750 µL PBS-BSA (final concentration 1 µg T4/µL).
4. Antithyroid drugs: 0.02% (w/v) 1-methyl-2-mercapto-imidazole (methimazole, MMI; Sigma, St. Louis, MO) and 1% (w/v) potassium perchlorate.
5. Surgical material: scissors, forceps, hemostatic forceps, clamps, 20–25 gauge (G) needles with blunted tips. The size of the needle used depends on animal age.
6. Anesthetics: ketamine and medetomidine.
7. Fixative: 4% (w/v) paraformaldehyde in 0.1 M PB, pH 7.2–7.4.
   The fixative should be freshly prepared in the fume hood the day of perfusion. For 1 L, heat 800 mL of water to 60–70°C, dissolve 40 g of paraformaldehyde on a magnetic stirrer while heating and slowly add a few drops (6–8) of 1 M NaOH. When the solution clears add 100 mL of 1 M PB, pH 7.2–7.4, add water to 1 L, filter, and use at room temperature (*see* **Note 1**).
8. 30% (w/v) Sucrose solution. For 250 mL dissolve 75 g sucrose in fixative at room temperature.
9. Tissue-TeK OCT compound and cryostat knife size 16 cm, type C.
10. Cryoprotective-buffered saline (CBS): 30% (v/v) ethylenglycol, 30% (v/v) glycerol, 40% (v/v) 0.1 M PB, pH 7.2.

2.2. In Situ *Hybridization*

Tubes, pipet tips, and glass material should be autoclaved. To prevent ribonuclease (RNase) contamination bake Pasteur and glass pipets for 2 h at 200°C and wear gloves at all times until the hybridization reaction has finished. Autoclaved high purity water is used in all steps. Diethyl pyrocarbonate (DEPC)-treated water is also used in some cases as indicated.

Table 1
Preparation of Stock Solutions and Hybridization Buffer

For a 45 mL stock solution, mix:	Final concentration in hybridization buffer:
25 mL Formamide (ultrapure)	50% (v/v)
5 g Dextran sulfate	10% (w/v)
5 mL of 50X Denhardt's	5X
6.25 mL of 5 M NaCl	0.62 M
1 mL of 1 M PIPES, pH 6.8	20 mM
1 mL of 0.5 M EDTA	10 mM
0.5 mL of 20% (w/v)	0.2 % (w/v)
sodium dodecyl sulfate (SDS)	

Add water to 45 mL. If SDS precipitates, warm the buffer to no more than 40°C.
Divide in 1.8-mL aliquots and keep frozen at –20°C.
This solution is valid for at least 3 mo.

Just before use, an aliquot is thawed and the following is added:	Final concentration in hybridization buffer:
100 µL of 1 M Dithiothreitol (DTT)	50 mM
50 µL of 10 mg/mL Salmon sperm DNA (ssDNA), boiled for 5 min and quenched in ice	250 µg/mL
50 µL of Yeast tRNA 10 mg/mL	250 µg/mL

1. TE, pH 8.0: to prepare 500 mL, mix 5 mL of 1 M Tris-HCl, pH 8.0, and 1 mL 0.5 M EDTA in a total volume of 500 mL water, pass the solution through a 0.45-µm filter and autoclave.
2. DEPC-treated water: prepare a 0.1% (v/v) solution of DEPC in a fume hood, tight close the bottle, shake the solution, heat at 37°C for 8 h, and autoclave to inactivate the DEPC.
3. Washing vials, multiwell plates, slide racks, and staining troughs.
4. PBS-Triton X-100. Triton X-100 is used to permeabilize the tissue. For minimum tissue damage, the final Triton X-100 concentration varies from 0.05 to 1% (v/v) depending on age: 0.05% for rat embryos and neonates, 0.1% for postnatal (P) d 5–15, 0.5% from P16 to adults, and 1% for old animals.
5. Triethanolamine solution: 0.25% (v/v) acetic anhydride in 0.1 M triethanolamine, pH 8.0. Prepare the triethanolamine in water and filter through a 0.22-µm filter. Take care because the acetic anhydride is very toxic and volatile, so it should be handled in a fume hood.
6. Hybridization buffer: prepare from stock solutions as indicated (**Table 1**) (*see* **Note 2**). For a 50-mL total working solution, a 45-mL partial solution is prepared

and kept frozen in 1.8-mL aliquots. Just before use, aliquots are thawed and the rest of components are freshly added.

7. Enzymes and other molecular biology reagents: restriction enzymes, RNA polymerases, RNase inhibitor, ribonucleotides, spin columns, proteinase K, DNA size markers, RNase A, and formamide, are of the highest analytical grade available.

8. [^{35}S]UTP, 12.5 mCi/mL, specific activity 1000–1500 Ci/mmol.

9. Stringency wash solutions:
   a. Solutions A, C, and D are prepared from a stock solution of 20X standard saline citrate (SSC) (NaCl 3 M, sodium citrate 0.3 M, pH 7.0) and β-Mercaptoethanol is added just before use.
   b. Solution A: 2X SSC, 10 mM β-Mercaptoethanol.
   c. Solution B: 50 mM Tris-HCl, pH 7.5, 5 mM EDTA, 0.5 M NaCl, and 4 µg/mL RNase A.
   d. Solution C: 0.5X SSC, 50% (v/v) formamide, 10 mM β-Mercaptoethanol. Formamide is added to diluted SSC.
   e. Solution D: 0.1X SSC, 10 mM β-Mercaptoethanol.

10. Solution E: 0.05% Triton-X100 in 0.01 M PB, pH 7.2.

11. Superfrost® microscope slides.

12. Graded ethanol solutions (50%, 70%, 90%) containing 0.3 M ammonium acetate when indicated.

13. Light-tight X-ray cassette and high resolution films.

14. Equipment needed for emulsion photography:
   a. Dark room.
   b. Two water baths set at 43°C.
   c. Special lamp with an appropriate safe-light filter. The regular red light used in most dark rooms cannot be used. If no special lamp is available, perform all procedures in complete darkness. We use a sodium lamp, commercially available from different sources, which gives good illumination and is safe.
   d. Anhydrous silica gel as desiccator.
   e. Humidifier.
   f. Thermometer.
   g. Photographic emulsion: Kodak NTB2, Ilford K-5 or Amersham LM-1.
   h. Dipping glass chambers, available from different providers: Amersham or Electron Microscopy Science.
   i. Black, light-tight, microscope slide archive boxes.
   j. Black electrical tape and plastic bags.
   k. Metacrilate boxes for housing the emulsion and the emulsion-coated slides.

15. Photographic developer and fix solutions: developer D-19 (Kodak) and AGEFIX (AGFA).

16. Richardson's blue staining solution: prepare a stock solution containing 0.5% (w/v) Azure II, 0.5% (w/v) methylene blue, and 0.5% (w/v) sodium borate in water, and then filter through paper. Before use, this stock solution is diluted 1/20 in 0.1 M PB, pH 7.2.

17. Blocking solution: 4% (w/v) BSA, 0.1% (v/v) Triton X-100, 5% (v/v) serum from the host of the secondary antibody, 0.1 *M* L-lysine in PBS, pH 7.2.
18. Antibodies: monoclonal or polyclonal primary antibodies specific for the proteins of interest and biotinylated secondary antibodies. Normal goat serum (NGS) and horse serum (HS).
19. Primary antibody solution: 4% (w/v) BSA, 0.1% (v/v) Triton X-100, 1% (v/v) serum in PBS, pH 7.2.
20. Secondary antibody solution: 4% (w/v) BSA, 0.1% (v/v) Triton X-100, 5% (v/v) serum in PBS, pH 7.2.
21. Avidin-biotin-complex (ABC) staining kit, available from different providers.
22. Diaminobenzidine (DAB) stock solution: 5 mg/mL in water, filter through 0.22-μm filter, aliquot, and store at –20°C.
23. Developer DAB solution: 0.5 mg/mL DAB in PBS, pH 7.2, plus 2.88 μL of a 35% H_2O_2/10 mL solution just before use (*see* **Note 3**).

3. Methods

3.1. Handling of Animals

3.1.1. Generation of Hypothyroid Rats and Mice

The induction of severe hypothyroidism is essential to observe changes in gene expression by thyroid hormone. Moderate hypothyroidism leads to physiological changes aimed at maintaining T3 concentrations in neural tissue within normal levels. The most important mechanism concerns deiodinase type 2 (D2). D2 is a selenoenzyme that catalyzes the removal of the iodine atom in the 5' position of T4 to generate the active hormone, T3. D2 activity is inhibited by T4 through a mechanism involving increased degradation of the enzyme in proteasomes. In situations of low T4, the increased expression and activity of D2, with the concomitant increased efficiency of T4 to T3 conversion, tends to maintain T3 concentrations constant *(30)*. Therefore, only under very severe hypothyroid conditions are T3 concentrations low in the brain, in contrast with other tissues, such as the liver or kidney.

Hypothyroidism can be induced in the rat by surgical or chemical means. Surgical thyroidectomy procedures are difficult in mice, owing to the anatomical features of the mouse thyroid gland, which is not easy to remove. In mice, as in rats, chemical thyroidectomy, combining two different antithyroid drugs gives satisfactory results.

3.1.1.1. SURGICAL THYROIDECTOMY

1. Anesthetize the rat.
2. Open a longitudinal incision in the neck.
3. Separate the borders of the skin and the neck muscles until the thyroid gland is exposed.

4. Pull off carefully each lobe of the gland.
5. Close the incision.

3.1.1.2. CHEMICAL THYROIDECTOMY

A number of drugs interfere with hormone synthesis in the thyroid gland. The most commonly used are sodium or potassium perchlorate, 1-methyl, 2-mercaptoimidazole (MMI) and 6-n-propylthiouracil (PTU). Perchlorate blocks the active transport of iodine to the thyroid gland carried out via the Na-I symporter. MMI and PTU block the intrathyroidal oxidation of iodine and, therefore, the iodination of thyroglobulin. PTU, in addition, inhibits the activity of type 1 deiodinase (D1), so that less T3 is formed from T4 in the liver and kidney. Since D2 is not inhibited by PTU, it is possible that D1 inhibition by PTU would spare T4 as a substrate for D2 in the brain. In our laboratory, we prefer the use of MMI for this reason and, also, because it is readily soluble in water. PTU, in addition to being less soluble, requiring alkalinization with NaOH, has a bitter taste and, therefore, is less tolerated by the animals. These drugs are given in the drinking water either alone or in combinations. In our laboratory, we use the following final concentrations: perchlorate: 10 g in 1 L of drinking water (1%); MMI: 200 mg in 1 L of water (0.02 %).

The most intense degree of postnatal hypothyroidism is achieved by giving MMI to the pregnant dams and then performing thyroidectomy of the newborns. Taking into account the ontogenic dates for the T3 receptor (E14) and the thyroid gland (E18), and that hypothyroidism does not interfere with placental metabolism during the second half of pregnancy, we usually administer MMI continuously to the dams in the drinking water, starting around d 9 to 10 postconception. Newborns are then surgically thyroidectomized at P5. Since MMI crosses the placenta, the newborns develop goiter, which facilitates removal of the gland. With this protocol, a severe hypothyroidism is induced. The drawback is the high mortality, which is due not only to the surgical stress, but also to the fact that hypothyroidism is so profound that few pups survive after weaning. To ensure longer viabiliy, the pups need to be kept with their mothers. As an alternative to surgery, the combination of MMI and perchlorate works well also and results in lower mortality.

3.1.2. Treatment with Thyroid Hormones

The hormones are administered to the animals by intraperitoneal injection, according to different schedules, such as:

a. Acute T3 dose: 50 µg of T_3/100 g body weight 24 h before killing.
b. Replacement dose of T3: 0.3 µg T_3 daily during at least 5 d before killing.
c. T4 treatment (replacement dose): 2 µg/100 g body weight.

3.2. Generating High Specific Activity RNA Probes

The riboprobe is synthesized from a 200–400 bp cDNA template from the gene of interest. Although longer probes may be used, this size range provides good diffusion into the tissue and good signal strength. Probes can be made from any region of the cDNA provided that there is no homology with related genes. The DNA template can be obtained by reverse-transcription polymerase chain reaction (RT-PCR) using specific primers, after which it is cloned in an appropriate vector containing promoters for phage RNA polymerases. Either sense or antisense RNA probes can be generated by linearizing the vector and using the appropriate polymerase. Use restriction enzymes leaving blunt or 5'-protruding ends to avoid nonspecific initiation of the polymerases.

The probe can be labeled with any radioactive ribonucleotide, but UTP is preferred because it works better empirically than other labeled ribonucleotides. As isotopes, ^{32}P, ^{33}P, or ^{35}S may be used. ^{35}S requires longer exposure times than ^{32}P, but gives more resolution in the autoradiographs. ^{33}P on the other hand requires shorter exposure times than ^{35}S, while producing better resolution than ^{32}P.

1. Prepare the transcription mix: add the following to a RNase-free microcentrifuge tube at room temperature: 1 µg of linearized proteinase K-treated DNA template; 2.5 µL of 10X transcription buffer provided with the polymerase (warm the buffer at room temperature to keep spermidine into solution); 1 µL each of 10 mM ATP, CTP, and GTP; 1 µL 0.75 M DTT; 40 U of RNase inhibitor; 50 µCi [^{35}S]UTP, 12.5 mCi/mL; 20 U of appropriate RNA polymerase, usually T7, T3, or SP6; DEPC water to a final volume of 25 µL.
2. Incubate 1 h at 37°C.
3. Add 50 µL TE, pH 8.0, and extract with phenol-chloroform 1:1 (v/v).
4. Remove the unincorporated nucleotides by gel filtration chromatography over a RNase-free column.
5. Measure the vol and count a 1-µL aliquot in scintillation fluid. Around 90 µL of total volume is recovered with an average $5–6 \times 10^5$ counts per min (cpm)/µL.
6. Add 1 µL of 1 M DTT per 50 µL of probe volume and store at –70°C if not used immediately.

3.3. Preparation of Brain Slices

3.3.1. Transcardiac Perfusion and Postfixation

1. Anesthetize the animal by intraperitoneal injection of a mixture of an anesthetic (ketamine 4 µg/100 g body weight) and an analgesic (medetomidine 10–15 µg/100 g body weight).
2. Place the animal on an operating table over a tray to collect the fixative while perfusing.
3. Cut open the thorax by a longitudinal incision and expose the heart.

4. Open the catheter and let the fixative to drip slowly while introducing the blunted needle tip carefully through the left ventricle wall and into the beginning of the aorta. Air bubbles are avoided by letting the fixative to run slowly just before the needle is inserted into the heart.
5. Cut open the right atrium and perfuse at a rate of 90–140 drops/min, depending on animal age. Perfuse with at least 3 mL of fixative per gram of body weight.
6. Open the cranium and carefully remove the whole brain in a piece, taking care not to damage the surface of the organ. Immerse the brain in fixative and postfix overnight at 4°C. Individual brains may be collected in 50-mL Falcon tubes containing the fixative.

3.3.2. Cryopreserving the Tissue

Immerse the brain in 30% sucrose solution and leave it undisturbed at 4°C until the brain sinks. This usually takes 2 to 3 d. At this point, the brain is ready to be frozen.

3.3.3. Freezing and Cutting the Tissue

1. Deposit a drop of OCT compound onto a piece of filter paper with the animal code written with pencil at the back.
2. Stick the cryoprotected brain to the paper in the desired orientation, and place the whole piece over powdered dry ice adding more on top of it until it is fully covered (*see* **Note 4**).
3. Leave for 10 min and store the frozen tissue at –70°C wrapped in transparent film.
4. To obtain the sections, allow the tissue to equilibrate at –20°C inside the cryostat for 20–30 min before sectioning, to prevent distorsion and fracturing of the tissue.
5. Cut the tissue at –20°C in 25-µm-thick sections. The sections are collected in 2-mL Eppendorf tubes (*see* **Note 5**) containing cold CBS and stored at –70°C. Sections are stable for years.

3.4. In Situ *Hybridization Using Radioactive Probes*

3.4.1. Tissue Processing

All incubations and washing steps should be performed under light shaking. The washing vials are previously cleaned with detergent and rinsed with water. This is followed by rinsing with ethanol, soaking in 3% H_2O_2 for 5–10 min at room temperature, and thoroughly rinsing with water.

1. Warm up the tube containing the frozen slices to room temperature, and place its content in a washing vial containing PBS. All procedures before prehybridization are performed at room temperature.
2. Rinse twice in PBS, for at least 5 min each.
3. Permeabilize the tissue by incubating the sections for 10 min in PBS containing Triton X-100. Triton concentration depends on animal age (*see* **Subheading 3.2.**, **item 4**).

4. Rinse with PBS for at least 5 min.
5. Deproteinize the tissue with 0.2 *N* HCl for 10 min.
6. Rinse with PBS for at least 5 min.
7. Acetylate sections with triethanolamine solution for 10 min to prevent nonspecific binding of the probe (*see* **Subheading 2.**).
8. Rinse with PBS for at least 5 min.
9. Postfix with 4% (w/v) paraformaldehyde in PBS for 10 min. This step is important in order to preserve the integrity of the tissue during hybridization reaction
   and stringency washes.
10. Rinse with PBS for at least 5 min and transfer the sections to 6-well plates.

3.4.2. Prehybridization and Hybridization Steps

These steps can be done in a water bath or in an oven.

1. Transfer the sections carefully, with the help of a bent Pasteur pipet, to a well of
   a 24-well plate containing 0.5 mL of hybridization buffer previously heated at
   55°C. Not more than 70 sections should be placed per well.
2. Incubate at 55°C for 3–5 h.
3. Heat the labeled riboprobe 10 min at 68°C and quench in ice.
4. Add the denatured riboprobe to the sections in the hybridization solution and
   incubate overnight at 55°C. Final concentration of the probe should be above
   10^7 cpm/mL. To avoid evaporation add PBS to the outer wells of the plate to
   make a humidity chamber. From this point on, it is not necessary to use RNase-free
   solutions and materials.

3.4.3. Stringency Washes and RNase Treatment

These steps help to remove the nonspecific binding of the probe.
All buffers should be preheated to the washing temperature.

1. First wash: transfer the sections from the 24-well plate to washing vials containing solution A and wash twice for 15 min at room temperature. The first wash is
   discarded as radioactive waste.
2. Second wash: 1 h at 37°C in solution B.
3. Third wash: twice for 1 h at 55°C in solution C.
4. Fourth wash: 1 h at 68°C in solution D.
5. Finally rinse thoroughly with PBS and then with solution E at room temperature.

The washed hybridized sections can be either mounted directly on microscope slides as described below, or used for immunohistochemistry before
mounting (*see* **Subheading 4.5.**).

3.4.4. Mounting the Sections

1. Place the sections in a Petri dish with solution E and deposit them flat to the
   surface of superfrost microscope slides (*see* **Note 6**).
2. Let the sections to air-dry for several hours.

3. Dehydrate the sections in graded ethanol solutions containing 0.3 M ammonium acetate to prevent loss of hybridizing RNA (sequentially, 2 min each in 50%, 70%, 90%, ethanol, and then 10 s in 100% ethanol).
4. Let the sections dry out under an air stream, for example in a fume hood, for 30 min.

3.4.5. Signal Detection

Signal detection is accomplished by autoradiography, either by apposing an X-ray film to the tissue sections for tissue level resolution, or by coating the sections directly with a photographic emulsion for cellular resolution. If the target mRNA is relatively abundant and does not need long exposure times, it is possible to perform both detection procedures in the same slide. If the probe needs long exposure times, it is better to perform both procedures in different slides. As a rule of thumb, it is not worth it to expose the sections for a longer time than the half-life of the isotope. Emulsion autoradiography needs around 3 to 5 times longer exposure than film autoradiography.

3.4.5.1. FILM AUTORADIOGRAPHY

1. The slides are apposed to sensitive films, for example Hyperfilm β-max® (Amersham), taking care that the sections contact the emulsion side of the film. The exposure time varies depending on the abundance of the target mRNA, and it may take from 3–70 d.
2. Develop the β-max films manually under safe light. It is very important that all fluids are at 20°C, especially the developer. The film is immersed in D-19 developer, with the emulsion side up, for 4 min under light shaking.
3. Wash the film with running tap water and fixed 5–10 min in fix solution diluted 1:7. It is finally washed under running water at around 20°C for at least 5 min.
4. The film is then air-dried.

The result of an experiment is illustrated in **Fig. 1**.

3.4.4.2. EMULSION AUTORADIOGRAPHY

This is a relatively simple procedure, but great care should be paid to many factors that can affect the final results. Because the extreme sensitivity of the emulsion to light and radioactivity, it is necessary to avoid any source of radiation in the darkroom and during the storage of the emulsion and the coated slides. The procedure must be done in complete darkness or under a safe light (*see* **Note 7**). The room should have an adequate degree of humidity, which can be achieved by running a humidifier before the procedure, if needed.

1. Heat the emulsion in a water bath at 43°C, for at least 15 min. Place the dipping glass chamber in another bath at the same temperature and fill it with the warmed emulsion. The dipping glass chamber should have been cleaned with diluted HCl, rinsed thoroughly with water, and dried.

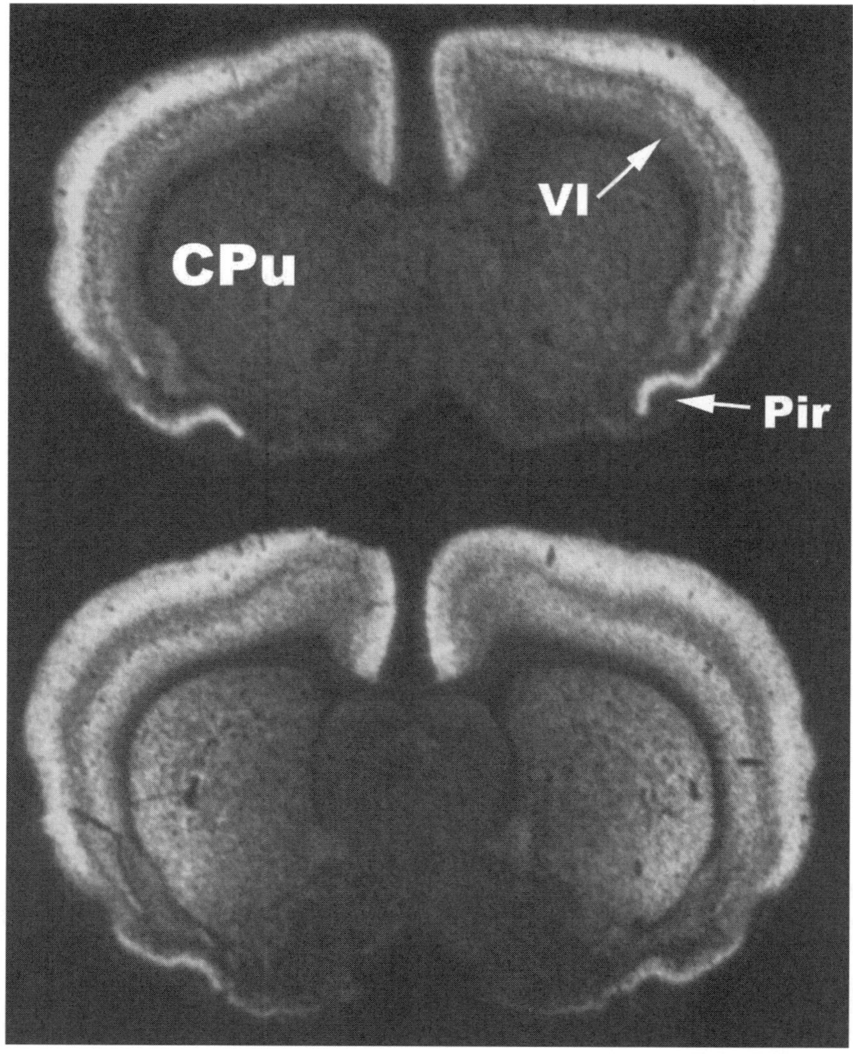

Fig. 1. Expression of RC3 in normal and hypothyroid rats. RC3 expression was studied by *in situ* hybridization using a $[^{35}S]$UTP-labeled riboprobe corresponding to the whole open reading frame. The results were examined by film autoradiography. To induce hypothyroidism, 0.02% MMI was giving in the drinking water to the pregnant dams, and the pups were thyroidectomized at P5. The hypothyroid and control rats were examined at P20. Upper panel: coronal section through the caudate of a hypothyroid rat. Lower panel: corresponding section of a normal control rat. Hypothyroidism induced a decreased RC3 expression in layer VI of the cortex and the caudate, and no changes in upper cortex layers or piriform cortex. Cpu, caudate-putamen; VI, layer VI of the cortex; Pir, piriform cortex. The data are from **ref. 23**.

2. Dip the slides, two at a time apposed by the backside, vertically into the emulsion for about 5 s. It is important to dip the slides vertical into the emulsion without lateral movements to obtain a homogeneous emulsion thickness.
3. Take them out of the chamber and blot the tip of the slides with Kimwipe.
4. Place them as vertically as possible and let them to dry out for 1–1.5 h.
5. Once dried, the slides are placed in a black slide box containing a small amount of desiccant wrapped in a Kimwipe. The position of the slides inside the box should be recorded. This helps in developing single slides to check the strength of the signal before processing the whole experiment.
6. The box is sealed and wraped in aluminum paper and placed in a metacrilate box at 4°C. The exposure time is variable, as indicated above.
7. To develop the slides, the box is brought at room temperature before opening to prevent moisture: 1 h should be enough.
8. Prepare 3 ethanol-washed glass slide troughs containing, respectively; 250 mL of D-19 developer diluted 1:1 with distilled water; 250 mL distilled water; 250 mL of AGEFIX diluted 1:7 with distilled water.
Cool the solutions down to 16°C. This can be accomplished by laying the troughs on top of ice.
9. Under safe light, place the slides sequentially, with intermittent mixing up and down, 3 min in D-19, 30 s in water, 5–10 min in fix solution, and then wash for 30 min in running tap water at around 20°C.
10. After the final wash in running tap water, stain for 2 min in Richardson's blue (*see* **Subheading 3.2.**, **item 16**) and rinse briefly in water the excess dye. Other staining solutions may be used provided they are not acidic.
11. Differentiate for approx 2 to 3 min in 70% ethanol, keeping control of the process to achieve the desired staining intensity.
12. Dehydrate in graded ethanol dilutions (80%, 96%, 100%, 100%) for 3 min each (*see* **Note 8**), followed by 3 min in 100% ethanol/xylene 1:1 (v/v), and then 5 min in xylene twice.
13. Coverslip the sections with DPX. The preparations are permanently stable if there are no air bubbles trapped between the coverslip and the tissue.

3.5. Double In situ *Hybridization and Immunohistochemistry*

3.5.1. Tissue Processing

After the washing steps from the hybridization reaction the floating sections in solution E, are ready for the immune reaction (*see* **Subheading 4.4.3.**).

3.5.2. Immune Reaction and Signal Detection with the ABC Method

Perform all the steps at room temperature under shaking except when indicated.

1. Incubate sections for 1 h in blocking solution (*see* **Subheading 3.2.**, **item 17**).
2. Wash 5 min with 0.1% Triton X-100 in PBS.
3. Incubate sections with the primary antibody diluted in primary antibody solution (*see* **Subheading 3.2.**, **item 19**, and **Note 9**).

4. Wash 3× with 0.1% Triton X-100 in PBS.
5. Incubate with the biotinylated secondary antibody diluted in secondary antibody solution (*see* **Subheading 3.2.**, **item 20**). The secondary antibody recognizes the primary antibody Fc region (*see* **Note 10**).
6. Before incubation time is over, prepare the ABC solution following the kit directions. This solution can be prepared about 30 min before use.
7. Wash 6×, 5 min each, with 0.1% Triton X-100 in PBS to completely remove the unbound secondary antibody.
8. Incubate the sections for 1 h in the ABC solution.
9. Wash twice for 5' with 0.1% Triton X-100 in PBS and then with PBS until foam disappears (it usually takes 4 washes of 5 min each).
10. Incubate the sections with developer DAB solution (*see* **Subheading 3.2.**, **item 23**). The developing time is variable and needs to be monitored under the microscope in each experiment.
11. Stop the reaction with PBS and wash several times for 5 min (*see* **Note 11**).
12. Mount the sections on microscope slides as described in **Subheading 4.4.4.** Perform film and emulsion autoradiography as described in **Subheading 4.4.5.**

A result of such an experiment is seen in **Fig. 2**.

3.6. Analysis of Results

All results may be analyzed in both qualitative and quantitative ways.

3.6.1. Autoradiograms

The X-ray films are scanned to generate digital images, which can be quantified with appropriate programs, such as National Institutes of Health (NIH) Image.

4.6.2. Microscopy Slides

Optical observation with light and dark field with a standard light microscope.

4. Notes

1. Paraformaldehyde is toxic, so the fixative should be prepared in a fume hood. It is very important to prepare correctly the fixative to obtain the best preservation of the tissue. Do not overheat the solution more than 60°C.
2. Use the sodium salt of PIPES, which is easier to dissolve than the free acid.
   50X Denhardt's is 1% (w/v) BSA, 1% (v/v) Ficoll 400 and 1% (w/v) polyvinylpyrrolidone. 20% SDS is prepared in water and autoclaved. The NaCl, PIPES, EDTA, and Denhardt's solutions must be filtered through a 0.22-μm filter before adding to the hybridization buffer.
   Dissolve the yeast tRNA in DEPC water. The ssDNA is dissolved in water by stirring in a beaker. It takes time to dissolve. Aliquots of both solutions are kept at –20°C.

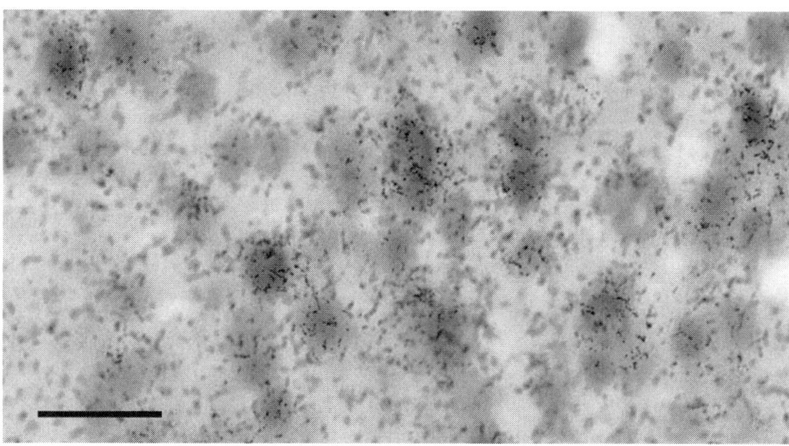

Fig. 2. Emulsion autoradiography using a labeled RC3 riboprobe combined with immunohistochemistry for T3 receptor β2 isoform in the rat retrosplenial cortex. Photography was carried out using a 63X immersion objective. Note the black silver grains over receptor positive cells. The scale bar equals 25 μm. Data are from **ref. 31**.

3. Take extra care because DAB is a carcinogen. All residues should be inactivated with bleach prior to discarding. It is also very hygroscopic, so the bottle should be warmed at room temperature before opening to avoid moisture.
4. The brain or brain blocks should be placed on the filter paper in such a way that the cutting plane is parallel to the paper base. For coronal sections, we routinely separate the cerebrum and the cerebellum by cutting perpendicularly at the level of the colliculi, and stick each part to the paper by the cutting plane. Brain blocks can be similarly prepared by cutting through different planes as desired.
5. For coronal sections, a whole adult rat brain can be collected into 3 series of 5 tubes each. The first set of tubes, labeled A1–A5 contains alternating serial sections from the olfactory bulb to the fimbria of the hippocampus. The second set, labeled B1–B5 contains the sections from the rest of the cerebrum. The third set, tubes C1–C5 contains the cerebellar sections. Thus, if one tube of each set is used, a whole representation of the brain is present in a single experiment. Each tube contains enough sections to be hybridized with 3 probes. In this way, one adult rat brain can be hybridized with 15 different probes.
6. If the slides are to be used for emulsion photography, the sections should be mounted in the lower 2/3 of the slide to ensure a uniform emulsion thickness.
7. Check that there is absolutely no light sources other than the safe light in the room where the dipping is being perform. Take care to cover all possible sources of light from the water baths with black tape, and avoid static electricity sparks. Advertise very clearly with signs outside the room that there is an experiment running to avoid anybody to go in.

8. The ethanol solutions used in these steps should not contain ammonium acetate because otherwise the stain gets out of the sections.
9. The temperature and the time of incubation depends on the specific antibody used and also on antibody dilution. We generally incubate overnight at 4°C, but in many cases, this has to be empirically determined.
10. Dilution of secondary antibody, incubation time, and temperature are specified in the ABC staining kit instructions, but sometimes has to be optimized empirically.
11. It is important to stop the reaction when the brown color resulting from DAB oxidation is still light, so that the hybridization grains after emulsion photography can still be seen over the immune background color.

Acknowledgments

We acknowledge the help provided by Dr. Luis de Lecea (Scripps research Institute, La Jolla, CA) in setting up the in situ hybridization and immunohistochemistry techniques described in this chapter. Work in our laboratory is supported by Grant No. PM98-0018 from DGICYT. A.G.-F. is supported by an investigator contract from the C.S.I.C., Spain.

References

1. Porterfield, S. P. and Hendrich, C. E. (1993) The role of thyroid hormones in prenatal and neonatal neurological development—current perspectives. *Endocr. Rev.* **14,** 94–106.
2. Legrand, J. (1984) *Effects of thyroid hormones on central nervous system, in Neurobehavioral Teratology* (Yanai, J., ed.), Elsevier Science Publishers, Amsterdam, pp. 331–363.
3. Pérez-Castillo, A., Bernal, J., Ferreiro, B., and Pans, T. (1985) The early ontogenesis of thyroid hormone receptor in the rat fetus. *Endocrinology* **117,** 2457–2461.
4. Bradley, D. J., Towle, H. C., and Young, W. S. (1992) Spatial and temporal expression of α- and β-thyroid hormone receptor mRNAs, including the β2-subtype, in the developing mammalian nervous system. *J. Neurosci.* **12,** 2288–2302.
5. Bernal, J. and Pekonen, F. (1984) Ontogenesis of the nuclear 3,5,3'-triiodothyronine receptor in the human fetal brain. *Endocrinology* **114,** 677–679.
6. Bernal, J. and Guadaño-Ferraz, A. (1998) Thyroid hormone and the development of the brain. *Curr. Op. Endocrinol. Diabetes* **5,** 296–302.
7. Rogister, B., Ben-Hur, T., and Dubois-Dalcq, M. (1999) From neural stem cells to myelinating oligodendrocytes. *Mol. Cell. Neurosci.* **14,** 287–300.
8. Sutcliffe, J. G. (1988) The genes for myelin revisited. *Trends Genet.* **4,** 211–213.
9. Vega-Núñez, E., Menéndez-Hurtado, A., Garesse, R., Santos, A., and Perez-Castillo, A. (1995) Thyroid hormone-regulated brain mitochondrial genes revealed by differential cDNA cloning. *J. Clin. Invest.* **96,** 893–899.
10. Alvarez-Dolado, M., Gonzalez-Moreno, M., Valencia, A., Zenke, M., Bernal, J., and Muñoz, A. (1999a) Identification of a mammalian homologue of the fungal Tom70 mitochondrial precursor protein import receptor as a thyroid hormone-regulated gene in specific brain regions. *J. Neurochem.* **73,** 2240–2249.

11. Iglesias, T., Caubín, J., Zaballos, A., Bernal, J., and Muñoz, A. (1995) Identification of the mitochondrial NADH dehydrogenase subunit 3 (ND3) as a thyroid hormone regulated gene by whole genome PCR analysis. *Biochem. Biophys. Res. Comm.* **210,** 995–1000.

12. Alvarez-Dolado, M., Iglesias, T., Rodríguez-Peña, A., Bernal, J., and Muñoz, A. (1994) Expression of neurotrophins and the trk family of neurotrophin receptors in normal and hypothyroid rat brain. *Mol. Brain Res.* **27,** 249–257.

13. Neveu, I. and Arenas, E. (1996) Neurotrophins promote the survival and development of neurons in the cerebellum of hypothyroid rats. *J. Cell Biol.* **133,** 631–646.

14. Aniello, F., Couchie, D., Gripois, D., and Nunez, J. (1991) Regulation of five tubulin isotypes by thyroid hormone during brain development. *J. Neurochem.* **57,** 1781–1786.

15. Aniello, F., Couchie, D., Bridoux, A. M., Gripois, D., and Nunez, J. (1991) Splicing of juvenile and adult tau mRNA variants is regulated by thyroid hormone. *Proc. Natl. Acad. Sci. USA* **88,** 4035–4039.

16. Ghorbel, M. T., Seugnet, I., Hadj-Sahraoui, N., Topilko, P., Levi, G., and Demeneix, B. (1999) Thyroid hormone effects on Krox-24 transcription in the post-natal mouse brain are developmentally regulated but are not correlated with mitosis. *Oncogene* **18,** 917–924.

17. Koibuchi, N. and Chin, W. W. (1998) *RORα* gene expression in the perinatal rat cerebellum: ontogeny and thyroid hormone regulation. *Endocrinology* **139,** 2335–2341.

18. Denver, R. J., Ouellet, L., Furling, D., Kobayashi, A., Fujii-Kuriyama, Y., and Puymirat, J. (1999) Basic transcription element-binding protein (BTEB) is a thyroid hormone-regulated gene in the developing central nervous system. Evidence for a role in neurite outgrowth. *J. Biol. Chem.* **274,** 23,128–23,134.

19. Cuadrado, A., Bernal, J., and Muñoz, A. (1999) Identification of the mammalian homolog of the splicing regulator *Suppressor-of-white-apricot* as a thyroid hormone regulated gene. *Mol. Brain. Res.* **71,** 332–340.

20. Alvarez-Dolado, M., Gonzalez-Sancho, J. M., Bernal, J., and Muñoz, A. (1998) Developmental expression of tenascin-C is altered by hypothyroidism in the rat brain. *Neuroscience* **84,** 309–322.

21. Alvarez-Dolado, M., Ruiz, M., del Rio, J. A., et al. (1999b) Thyroid hormone regulates reelin and dab1 expression during brain development. *J. Neurosci.* **19,** 6979–6973.

22. Alvarez-Dolado, M., Cuadrado, A., Navarro-Yubero, C., et al. (2000) Regulation of the L1 cell adhesion molecule by thyroid hormone in the developing brain. *Mol. Cell. Neurobiol.* **16,** 499–514.

23. Iñiguez, M. A., De Lecea, L., Guadaño-Ferraz, A., et al. (1996) Cell-specific effects of thyroid hormone on RC3/neurogranin expression in rat brain. *Endocrinology* **137,** 1032–1041.

24. Krebs, J. and Honegger, P. (1996) Calmodulin kinase IV: expression and function during rat brain development. *Biochim. Biophys. Acta* **1313,** 217–222.

25. García-Fernández, L. F., Urade, Y., Hayaishi, O., Bernal, J., and Muñoz, A. (1998) Identification of a thyroid hormone response element in the promoter region of

the rat lipocalin-type prostaglandin D synthase (beta-trace) gene. *Mol. Brain Res.* **55,** 321–330.

26. Falk, J. D., Vargiu, P., Foye, P. E., et al. (1999) Rhes: a striatal-specific Ras homolog related to Dexras1. *J. Neurosci. Res.* **57,** 782–788.

27. Valentino, K. L., Eberwine, J. H., and Barchas, J. D. (1987) *In Situ Hybridization: Applications to Neurobiology*, Oxford University Press, New York.

28. Wisden, W. and Morris, B. J. (1994) *In Situ Hybridization Protocols for the Brain*, Academic Press, London.

29. Polak, J. M. and McGee, J. O. D. (1998) *In Situ Hybridization: Principles and Practice*, Oxford University Press, Oxford.

30. Obregón, M. J., Ruiz de Oña, C., Calvo, R., Escobar del Rey, F., and Morreale de Escobar, G. (1991) Outer ring iodothyronine deiodinases and thyroid hormone economy: responses to iodine deficiency in the rat fetus and neonate. *Endocrinology* **129,** 2663–2673.

31. Guadaño-Ferraz, A., Escámez, M. J., Morte, B., Vargiu, P., and Bernal, J. (1997) Transcriptional induction of RC3/neurogranin by thyroid hormone: differential neuronal sensitivity is not correlated with thyroid hormone receptor distribution in the brain. *Mol. Brain Res.* **49,** 37–44.

6

The v-*erbA* Oncogene

Assessing Its Differentiation-Blocking Ability Using Normal Chicken Erythrocytic Progenitor Cells

Olivier Gandrillon

1. Introduction

The v-*erbA* oncogene is the most clear-cut case of an oncogene that acts by blocking a differentiation sequence (for a review, *see* **ref. *1***). It has been discovered as one of the two oncogenes carried by the avian erythroblastosis virus (AEV), a leukemia-inducing retrovirus. It is derived from the c-*erbA* proto-oncogene, which encodes the α form of the nuclear receptor for the thyroid hormone triiodothyronine (T3Rα) (*2,3*; for a detailed description of the structural differences between v-erbA and T3Rα, *see* **ref. *4***). Thyroid hormone receptors belong to the large superfamily of structurally and functionally related receptors that includes the receptors for thyroid hormone (T3R), retinoic acid (either all-*trans* or 9-*cis* isoforms, RARs and RXRs) and vitamin D3 (VD3R), which are all lipophilic ligand-regulated transcription factors (*5*).

The v-erbA oncoprotein is widely thought to function as an antagonist of normal T3R and related receptors in the control of transcription (*6,7*). Functional and biochemical dissections of this oncogene have brought many informations on the mechanisms of action of normal receptors and on the ways through which altered receptors can contribute to oncogenic transformation (*8*; for a review, *see* **ref. *4***).

The precise mechanism of action of v-erbA is nevertheless still unclear. The most widely accepted hypothesis at the present time is that v-erbA behaves as a constitutive T3 apo-receptor, maintaining the silencing of T3-responsive genes in the presence of T3 (*9,10*). Such a constitutive silencing is thought to result from the constitutive recruitment of corepressors like SMRT or NCoR (*11,12*). A co-hypothesis has been that this silencing function might extend to

From: *Methods in Molecular Biology, Vol. 202: Thyroid Hormone Receptors: Methods and Protocols*
Edited by: A. Baniahmad © Humana Press Inc., Totowa, NJ

the normal functioning of RARs *(7)*. It has also been shown that T3R, RAR, and RXR can modulate gene expression through an indirect way by inactivating the AP-1 transcription factor in a ligand-dependent fashion *(13–15)*. In that case as well, the v-erbA oncogene was shown to transdominantly abrogates the ability of T3R and RAR to inactivate AP-1 *(13)*.

Although the initial model, stating that v-erbA acts by its dominant negative activities toward nuclear hormone receptors is supported by numerous experimental evidences *(16–18)*, it is nevertheless unable to accommodate a wealth of recent reports demonstrating the existence of subtle differences between the unliganded T3Rα and the v-erbA oncoprotein, in terms of subcellular localization *(19)*, translational regulation *(20)*, competition for a transcriptional repressor *(21)*, binding of cofactors *(8,22)* and DNA binding specificity *(23–25)*. In any case, the validation of this hypothesis still awaits the identification of the relevant v-erbA target genes in a v-erbA target cell.

The v-erbA oncogene acting alone (i.e., in the absence of an other oncogene) has been shown to modify the biology of a wide variety of cells. This includes altering the growth of chicken embryo fibroblasts (CEFs) *(26)*, chicken embryo neuroretina cells (CNRs)*(27)*, glial cells *(28,29)*, and the differentiation potential of PC12 cells *(30)*, avian myoblasts *(31)* and erythroblasts *(see below, this section)*. Furthermore, v-erbA has been shown to cooperate with viral oncogenes v-*erbB* *(see below, this section)*, v-src, v-fps, v-sea, v-Ha ras, v-myc and v-ets *(27,32–34)* as well as with the normal signaling elicited through epidermal growth factor (EGF) receptor or bFGF receptor *(27,35)*.

The generation of mice expressing the v-erbA oncogene as an ubiquituously expressed transgene has paved the way for the exploration of v-erbA action in a wide variety of mammalian tissues. The transgenic mice indeed displayed very numerous abnormalities including hypothyroidism, reduced fertility, decreased body weight due to reduction in adipose tissue, and sex-specific defects: transgenic females displayed behavioral abnormalities by abandoning or consuming their pups, whereas transgenic males developed seminal vesicle abnormalities and hepatocellular carcinomas *(36)*. This generation of carcinomas in vivo definitely established v-erbA as a bona fide oncogene.

The most well studied biological function of v-erbA to date is the function it fulfills within its natural host, the retrovirus AEV. When injected intravenously into newborn chicks, AEV induces acute and fatal erythroleukemia. When injected subcutaneously, it also induces sarcomas, but the erythroleukemia usually quickly takes over and kills the animal before the sarcoma has had time to extend. This observation could also explain that no other pathology has yet been described in chickens infected by AEV.

AEV carries the v-*erbA* and the v-*erbB* oncogenes. Both oncogenes cooperates through direct and indirect mechanisms (**Fig. 1**).

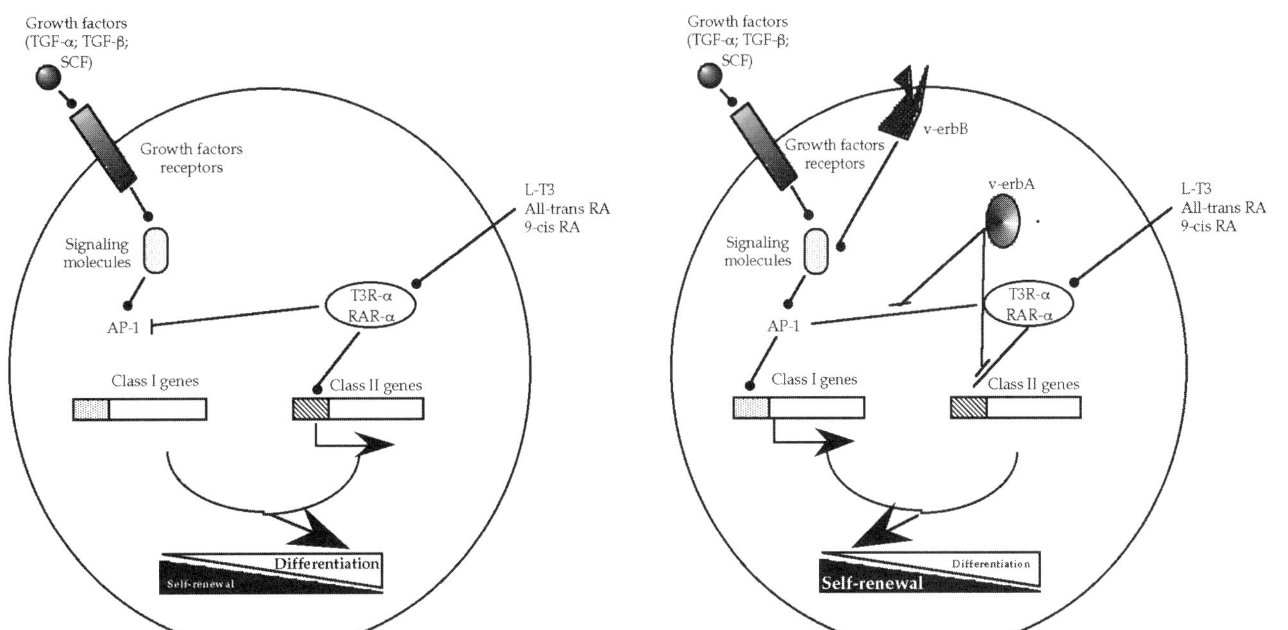

Fig. 1. Control of gene expression by c-erbA, RAR, and AP-1 (left panel) and its deregulation under the influence of v-erbA and v-erbB oncogenes (right panel). The arrows with solid circles denote an activation, and the arrows with vertical bars denote a repression mechanism. The left panel shows the normal pattern of gene regulation in cells treated with T3 or RA. Class I genes represent AP-1 dependent genes under the control of TPA-responsive element (TRE, vertical hatch marks). Class II genes represent hormone-regulated genes under the control of T3RE or RARE (diagonal hatch marks). This progenitor cell will switch from a self-renewal program into a differentiation program. The right panel depicts the pattern of gene expression in v-erbA- and v-verbB-expressing cells. This transformed cell will be blocked in an actively proliferating self-renewing state.

The v-erbB oncogene, which is a mutated transforming growth factor-α receptor (TGF-α-R) endowed with constitutive tyrosine kinase activity *(37)*, only induces an hyper proliferation of immature erythroblasts, presumably by the constitutive activation of TGF-α-R controlled signaling, although this still has to be formally demonstrated in erythroblasts. This hyperproliferative response causes only a partial differentiation arrest. The transformation promoting activity of v-erbA in AEV is widely thought to lie in its ability to arrest the differentiation of v-erbB-expressing erythroblasts by effectively silencing stage-specific erythroid genes such as the carbonic anhydrase II gene (CA II) that are otherwise activated by liganded T3R or RAR *(38–40)*. It has indeed been demonstrated that both T3Rα and RARα receptors do play a key role in switching erythrocytic progenitor cells out of the self renewal compartment and making them enter a new genetic program leading the cells to either die by apoptosis or differentiate into a fully mature erythrocyte *(18,41)*. This key role of nuclear hormone receptors have been recently strengthened by the description of the erythrocytic blockade appearing in NCoR$^{-/-}$ knock-out mice *(42)*. In this case, the premature up-regulation of genes normally inhibited by the aporeceptors seems to be responsible for the observed erythroid differentiation failure beyond the bust-forming unit erythroid (BFU-E) stage.

The v-erbA oncogene might then act by its dominant negative activity toward nuclear hormone receptors and prevent the exit from the self-renewal compartment by blocking the T3R/RAR-controlled gene activation (*see* **Fig. 1**). This hypothesis remain, nevertheless highly speculative as long as the identity of the relevant target genes involved are, for the most part, unknown.

One main reason for this paucity of relevant information concerning the v-erbA mode of action relates in part to the lack of a cellular system, in which the v-erbA activity can be assessed that would fulfill all of the following criteria at the same time:

1. To allow the maintenance of the most immature progenitors in a self-renewing state, while allowing to induce those cells at any time into a complete differentiation process. This would allow v-erbA action to be investigated in its natural cell target.
2. To be based on normal cells taken directly from live animals without any genetic alterations. This would allow to investigate v-erbA activity to be assessed in the absence of any interference (like a residual kinase activity in the systems employing temperature-sensitive kinases *[43]*).
3. To be easily manipulated, so as to allow its transplantation in various laboratories. This would allow experimental data to be replicated.
4. To be easily scaled up, so as to permit molecular biology and biochemical analyses. This would be absolutely needed in order to assess the biological activities and the molecular actions within the very same cells.

We have recently devised a model system that fulfills all of the above. This results from the cooperative action of TGF-α and TGF-β on chicken erythroid progenitor self-renewal *(44)* generating the outgrowth of erythrocytic stem cells, which we called T2ECs, for TGF-α/TGF-β-induced erythrocytic cells. We will describe it in detail in the following section with a special emphasis on the manner this system can be used for assessing the differentiation-blocking activities of the v-erbA or any other oncogene.

2. Materials

1. β-Mercaptoethanol : prepare freshly every week as a 10^{-1} M solution in α-modified Eagle's medium (MEM) starting from a 14 M stock solution. For stability issues of all solutions, (*see* **Notes 1–3**).
2. Dexamethasone: purchased from Sigma as a cell culture tested powder; prepare as a 10^{-3} M stock solution in ethanol.
3. TGF-β: purchased from various sources (R&D, Promega, TEBU) as a human recombinant protein; purity > 97% as determined by sodium dodecyl sulfate polyacrylamide gel electrophoresis (SDS-PAGE); prepare as a 1 μg/mL stock solution in 0.05 M HCl/1% fetal bovine serum (FBS).
4. TGF-α: purchased from various sources (R&D, Promega, TEBU) as a human recombinant protein; purity > 97% as determined by SDS-PAGE; prepare as a 10 μg/mL stock solution in 10 mM acetic acid/1% FBS.
5. LM1: for 100 mL of medium, mix in the following order: 89.2 mL α-MEM 10 mL FBS (selected batch), 1 mL penicillin plus streptomycin; 100 μL of 1 M Hepes; 100 μL of 10^{-1} M β-Mercaptoethanol; 100 μL 10^{-3} M dexamethasone; 100 μL 10 μg/mL TGF-α ; 100 μL 1 μg/mL TGF-β.
6. LM 23: for 10 mL of medium, mix in the following order: 10 mL α-MEM; 10 μL 1 M HEPES; 10 μL 10^{-1} M β-Mercaptoethanol; 10 μL 10^{-3} M dexamethasone; 10 μL 10 μg/mL TGF-α , 10 μL 1 μg/mL TGF-β (**Note 1**).
7. G418: purchased from Roche; purity > 80% (LTC): prepare a 100 mg/mL stock solution by dissolving 1 g of G418 under 10 mL of α-MEM containing 150 μL of 10 N NaOH, in order to neutralize the acidic function of G418.
8. 3% Phenylhydrazine solution: mix 450 μL phenylhydrazine, 98%, plus 15 mL of phosphate-buffered saline (PBS) 1X.
9. DM17: for 10 mL of medium, mix in the following order: 7.4 mL α-MEM; 1 mL FBS; 1.5 mL anemic chicken serum (ACS) (*see* **Subheading 3.3.1.**), 100 μL penicillin plus streptomycin; 10 μL of 1 M HEPES; 10 μL 10^{-1} M β-Mercaptoethanol; 50 μL 0.115 U/mL insulin.
10. Insulin: dissolve 10 mg of insulin (from bovine pancreas; CC-tested; equivalent to 500 USP U/mg) under 1 mL of 0.02 M HCl. Filter using 0.22-μm filters, aliquot trough 20 μL. Prepare the working solution by dissolving the frozen 20-μL pellet directly in 415 μL of α-MEM plus 10% FBS. Do not reuse.
11. PBSA: add 1 mL of a 100 mg/mL bovine serum albumin (BSA) solution (purchased from Roche) to 99 mL of PBS.

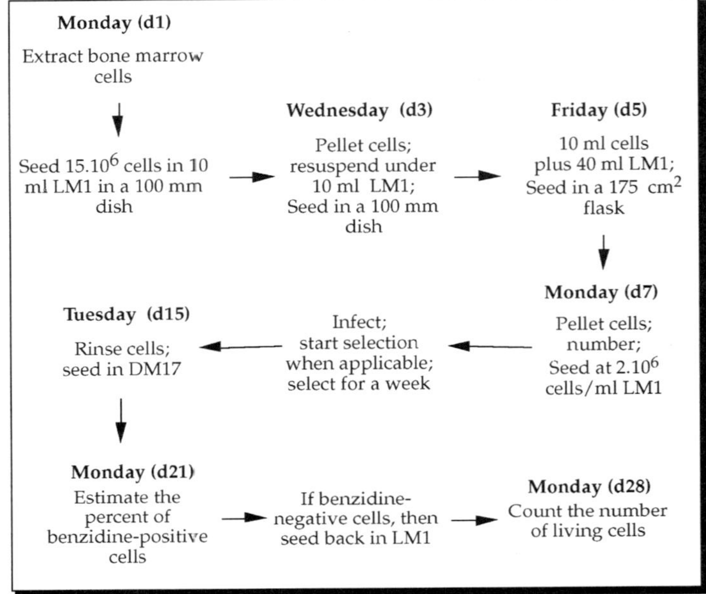

Fig. 2. A general outline of the overall procedure generating T2ECs

12. Acetic solution of benzidine: resuspend 100 mg of benzidine (benzidine dihydrochloride Isopac; Sigma, cat. no. B0386) under 50 mL of an 3% acetic acid solution (1.5 mL acetic acid plus 48.5 mL water). Resuspend the whole content of the Isopac bottle by injecting through the butyl rubber stopper. Store at room temperature protected from light.

13. The Ro41-5253 molecule *(45)* (a kind gift of Dr. M. Klaus [Hoffmann-La Roche Ltd.; Basel] is prepared as a $10^{-4} M$ stock solution in dimethyl sulfoxide (DMSO).

3. Methods

3.1. Generating T2ECs

A general outline of the overall procedure is represented in **Fig. 2**.

3.1.1. Obtaining Bone Marrow Cells

1. Procure 19-d-old chicken embryonated eggs. We usually use SPAFAS white leghorn, but we have verified that other strains (including commercially available, outbred chicken flock *[44]*) can also be used, as long as their ability to be infected by the desired retroviruses (cf 2.1) is ascertained.

2. Remove shell from air sac end with sterile forceps. Pour content (embryo plus yolk sac) on a paper sheet on top of an aluminum foil. Cut embryo's head, and rinse the embryo's body with 70% alcohol.

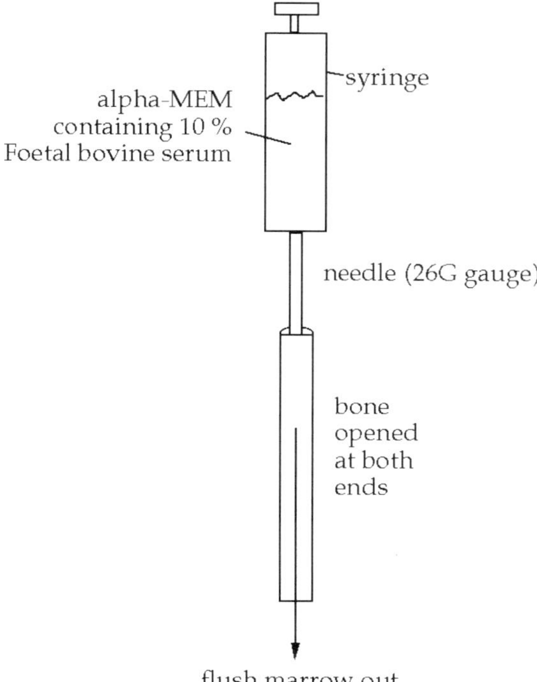

Fig. 3. Obtaining bone marrow cells from chicken tibias and flushing out the marrow of the bone. For details *see text*.

3. Place the embryo on a fresh paper sheet on top of an aluminum foil, take under a sterile hood.
4. With sterile scissors, cut open both chicken thighs. Dissect the tibias out of the flesh and muscles by cutting the bone at both ends (one should try not to make a frank cut, but more a "cut while breaking" so as to take out the bone as clean from soft tissue as possible).
5. Flush out the marrow out of the bone with α-MEM containing 10% FBS (α-FBS), using a 26-gauge needle, in a sterile plastic can. Use about 1 mL of α-FBS per tibia (**Fig. 3**).
6. Change the needle to a 18-gauge, and dissociate the cells by flushing them strongly in and out of the syringe about 10 times, in the sterile can. Transfer to a plastic tube suited for centrifugation .
7. Spin cells at 350g for 5 min at 4°C. Resuspend in LM1 (*see* **Subheading 2.**), number the cells and seed in LM1 at 1.5×10^6 cells/mL. In our experience, the cells will prefer to be grown under small volumes (500 µL per well of a 24-well dish is fine), but larger flasks are also suitable (*see* **Subheading 3.1.2.**). No plastic- or brand-related specificity has ever been evidenced (*see* **Note 1**).

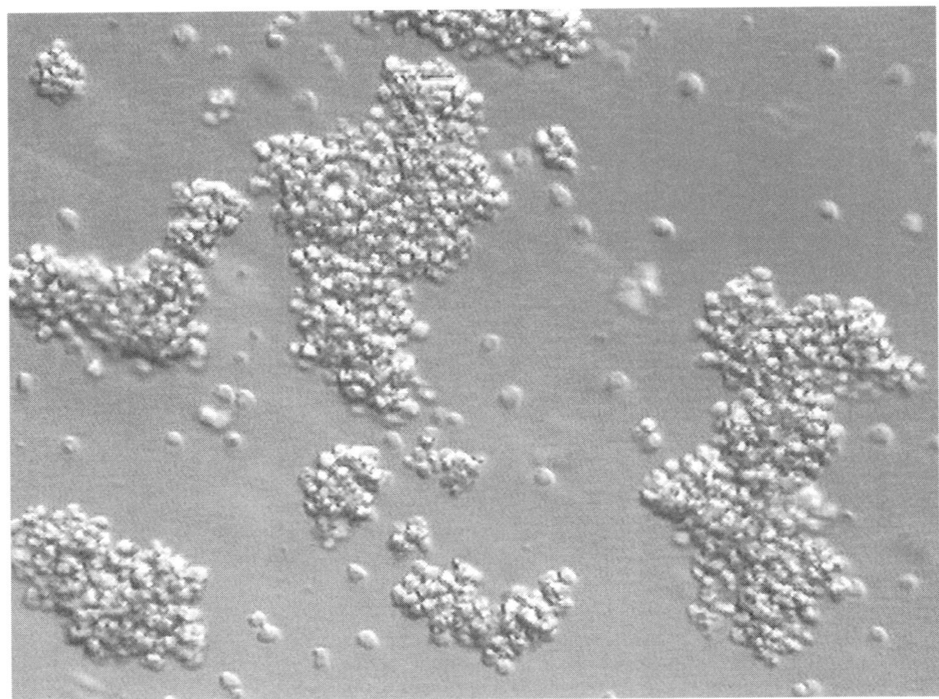

Fig. 4. Phase-contrast microphotography of living T2ECs.

3.1.2. Growth Conditions

The exact pattern of medium changes and cell passages will, of course, depend upon local growth conditions. Remember that the first week in culture is the critical period, and if you do not obtain extremely large tightly packed clumps of nonadherent round and very refractile cells (**Fig. 4** and [http:// cgmc.univ-lyon1.fr/Gandrillon/T2ECs.jpg]) after that first week, chances are that your culture will not start. In that case, restart a culture taking special care of using fresh factors (*see* **Note 1**).

We usually conform to the following pattern (see general outline of the overall procedure upper): bone marrow is prepared on Monday, and 15×10^6 cells are seeded in 10 mL of LM1. Those cells are centrifuged on Wednesday and resuspended in 10 mL LM1 and passed on Friday (10 mL cells plus 40 mL LM1 in a fresh 175-cm^2 flask). On the following Monday, cells are pelleted, numbered, and seeded at a density of about 2×10^5 cells/mL. Cells are then passaged every other day, or so, using a 1/5 to 1/10 dilution ratio.

3.2. Infecting T2ECs

3.2.1. Constructing the Viral Vector

It is beyond the scope of this chapter to go into extensive details of how to construct a proper retroviral vector expressing the gene of interest. There are mainly two strategies that can be envisioned:

1. Simply mimic the situation that is observed for v-erbA in the original AEV: that is to fuse the gene of interest in frame with the residual gag sequence. A pXN22 vector (a very close relative of the described pTXN3' *[46]*) can be obtained at will from the author. At least three cloning sites within the gag sequence (*Xho*I, *Xba*I and *Apa*I) can be used for in-frame cloning. This will generate a neo^R-expressing defective retroviral vector. It then will have to be cotransfected with a viral helper like RAV-1 in order to generate a replication-competent viral stock.
2. An alternative strategy consists in using the replication competent RCAS virus. RCAS is a virus that is derived from the Rous sarcoma virus. It has the three genes needed for viral replication (gag, pol, env) plus an enginered ClaI unique cloning site *(47)*. In that case, the whole cDNA of interest (including its own ATG codon) has to be inserted *(48)*.

3.2.2. Generating Viral Stocks

In order to generate viral stocks, the viral recombinant genomes will have to be transfected either in CEFs *(46)* or in T2ECs. Only this second technique will be described here.

1. Prepare the following solutions in 2 Eppendorf tubes: Solution A. 25 µL of LM23 (serum-free antibiotics-free medium, *see* **Subheading 2.**) plus 4 µL of Lipofectamine™; and Solution B. 25 µL of LM23 plus 0.7 µg of endotoxin-free plasmid DNA (we routinely use the EndoFree™ Plasmid Kit from Qiagen to prepare our plasmid solutions). In the case where a defective genome has to be cotransfected with a helper genome, use a defective/helper ratio of 5:10, for a total amount of DNA of no more than 1 µg.
2. Combine solutions A and B together, mix gently, and incubate at room temperature for 25 min.
3. Centrifuge exponentially growing T2ECs (use a more than a 1-wk-old culture), resuspend them under LM23, number the cells, adjust the concentration to 10×10^6 cells/mL. Seed 200 µL of cells/well of a 24-well dish.
4. Add gently the DNA/Lipofectamine™ mixture to the cells. Incubate at 37°C for 1 h.
5. Add 250 µL of LM23 plus 20% FBS. Incubate overnight.
6. Start the applicable selection (*see* **Note 2**). If no selection is applicable, grow the cells for about 10 d, at which time the infection should be maximal. During that time period, cells should grow normally and, therefore, should be diluted every other day or so. Verify the expression of your protein under scrutiny by conventional methods.

7. Rescue supernatant by centrifugation at 1500*g* for 15 min, sterilize using a 0.22-µm filter, store aliquoted at –80°C. Do not reuse a thawed vial (*see* **Note 3**).

3.2.3. Titrating Viral Stocks

1. If G418 selection is applicable, then titer the viral stocks by infecting CEFs with serial dilutions of the viral stock, grow cells in the presence of 200 µg/mL G418, and count the number of resistant-forming units (rfu) 15 d after infection, as described previously *(26)*.
2. If the virus carries a gene able to transform CEFs, the titer is determined as colony forming units (cfu) in the transforming focus assay *(50)* after infecting CEFs with serial dilutions of the viral stock.
3. If neither condition 1 nor 2 is met, then the titer has to be estimated through a p27 expression assay using QT6 cells, which is a quail fibroblast cell line *(51)* grown in the same medium as CEFs. For titer estimation, QT6 culture are infected with 0.1, 1, 10, and 100 µL of viral stock, and the titer is determined 12 h later by an immunoassay against the viral p27^{gag} protein. Briefly, infected cells are fixed at 4°C with formaldehyde 4% in PBS, rinsed twice with PBS, and once with Triton 0.25% in PBS and incubated with a rabbit anti-p27^{gag} antibody for 1 hour. Cells are rinsed twice with Triton 0.25% in PBS and incubated with an alkaline phosphatase-conjugated anti-rabbit IgG antibody for 1 h. Cells are then rinsed twice with PBS and incubated with the alkaline phosphatase substrates Naphtol ASMX Phosphate (0.4 µg/mL) and Fast Red TR Salt (6 µg/mL in Tris 0.1 *M*; pH 8.3). Finally, the cells are rinsed with water, and the colonies containing red-stained cells are numbered.

3.2.4. Infection

Resuspend 10^6 T2ECs (use a more than a 1-wk-old culture) under 500 µL of viral suspension. Seed in a 24-well dish. The next day, dilute by 2 mL of LM1 and start the applicable selection procedure (*see* **Note 2**). Remember to always carry at all times cells infected with a control retrovirus, either an empty retrovirus or a retrovirus carrying a nonactive form of the protein under scrutiny. As far as v-erbA is concerned, we routinely use two controls: a retrovirus harboring the *neo*^{R} gene only (XJ15) *(26)*, and a retrovirus harboring the v-erbAS61G mutant, a nontransforming point mutant of v-erbA *(7)*.

3.3. Test for the Differentiation-Blocking Ability

3.3.1. Preparing the ACS Through Phenylhydrazine Injections

The ACS is a mandatory component of the differentiation medium. It provides cells with an optimally balanced differentiation factors combination, containing most likely erythropoietin, but also nondefined cytokines *(49)*.

1. Day 1: Inject adult chicken (more that 1-mo-old) intraperitoneally with 60 µL of a 3 % phenylhydrazine solution per 100 g of chicken weight.

2. Day 2: Inject intraperitoneally with 75 µL of a 3% phenylhydrazine solution per 100 g of chicken weight.
3. Day 3: Put the chicken on a complete diet (nothing to eat).
4. Day 4: Anesthetize birds through injections of a combination of ketamine mixed with xylazine (25–30 mg/kg intravenous [IV] or intramuscular [IM]). Use a minimal dose (1/4 of calculated dose) and repeat after 5–10 min the minimal dose until the desired plane of anesthesia is reached.
5. Collect blood through heart puncture. Kill the birds without letting them regaining consciousness through CO_2 gas overdose.
6. Let the blood clot in an Erlenmeyer flask overnight at room temperature.
7. Day 5: Dilacerate the clot and centrifugate at 2500*g* for 10 min at 4°C. Rescue supernatant, filter on 0.45-µm membrane, and store aliquoted at –80°C.

3.3.2. Inducing and Assessing Differentiation

1. Centrifuge exponentially growing T2ECs (use a more than a 1-wk-old culture), resuspend them under α-MEM plus 10% FBS, number the cells, centrifuge and resuspend at a concentration of 10^6 cells/mL of DM17 (*see* **Subheading 2.**). Seed 1 mL of cells/dish of a 12-well dish. For a representative kinetics of differentiation in that system, *see* **ref. *41***. Maximum differentiation is obtained after 6 d.
2. The level of differentiation is estimated by acidic benzidine staining.
   For this, prepare the working benzidine solution extemporaneously: take the needed amount (25 µL/experimental point) of acetic benzidine solution (*see* **Subheading 2**, **item 12**) through the butyl rubber stopper and add 20 µL of H_2O_2 (30% [w/w], Sigma)/mL of benzidine solution. Mix and keep away from light. Discard within 1 h after preparation.
   Then, 100 µL of the cell suspension is taken from each well, added on top of 800 µL of PBSA, centrifuged at 4000*g* for 5 min at 4°C, and the pellet is resuspended in 25 µL PBSA. Then add 25 µL of benzidine-H_2O_2 solution, mix, and keep at room temperature away from light for 5 min. Add on a Kova® slide (Hycor), and count the percent of benzidine-positive (= blue) cells on at least 100 cells per point. A convenient control is made of cells maintained in LM1 medium, where the percentage of positive cells should be very low (in the 1% range).

3.3.3. Testing for Differentiation Blockade

Infected cells are first selected, when applicable (*see* **Note 2**), and induced to differentiate for 6 d.

The differentiation can be followed during that time period on a daily basis, but this is not an absolute requirement. Note also that no obvious difference might be apparent for the early differentiation period.

After 6 d of differentiation, estimate the percent of benzidine positive cells. At that point, control noninfected cells, or cells infected with control viruses should display at least 95% of benzidine-positive cells. Any well displaying a lower percentage of positive cells should be considered a strong indication of a differentiation blockade, especially in those cultures where the benzidine-negative cells display the classical morphology of very round refringent actively growing T2ECs.

In order to definitely assess the differentiation block, simply rinse the cells and seed them back in LM1 medium. No cells whatsoever should grow out of control cultures in those conditions, and only those cells that escaped the growth inhibition associated with erythroid terminal differentiation will be able to grow. A very simple growth curve will then be sufficient to document the phenomenon.

The stringency of the differentiation blockade can be assessed from: (*i*) The frequency of benzidine-negative cells in the first differentiation assay; and (*ii*) The reinduction of blocked cells to redifferentiate (i.e., cells grown in LM1, then DM17, then LM1, then DM17 once more).

3.3.4. Assessing the Role of Normal Receptors in the Differentiation Process

Most of the factors (peptide hormones, steroids, vitamins, etc.) that have been shown to affect the erythroid differentiation process are present at trace level in numerous biological fluids (mainly sera) used in various culture systems. Furthermore, some of these factors are produced in an autocrine fashion. Therefore, the role of these factors usually has to be analyzed only in defined systems specifically deprived in a given factor (using specific inhibitors, depleted sera, neutralizing antibodies, as well as dominant-negative isoforms).

Furthermore, most of the effects we observed relied upon a cooperative action of at least two factors. Studying one type of factor without controlling the effect of the other might result in pretty diverse biological readouts.

In the case of thyroid hormone action, the final potential source for confusion stems from redundancy. Indeed, we have shown that the potential sources of hormones activating both T3Rα and RARα have to be neutralized in order to obtain a significant decrease in the differentiation ability of the progenitors *(41)*. Such a co-depletion was accomplished by preincubating the complete differentiation medium with the anti-T3 (monoclonal antibody from BioGenesis) at 0.3% (v/v) and the RARα-specific antagonist Ro41-5253 at 10^{-7} *M* at 4°C for 1 h under constant agitation. Under those conditions, a 40% decrease in the percent of benzidine-positive cells should be observed *(41)*.

4. Notes

1. The freshness of all LM1 components will be the decisive factor for generating healthy rapidly growing T2ECS cultures. Almost all of the components used can suffer from excessive aging. This includes:
   a. α-MEM: do not use a bottle of α-MEM that has been opened for too long (more than about 2 wk), as indicated by its strongly pink color.

b. FBS: after being decomplemented, aliquot, and store at –20°C. The exact nature of the serum batch to use seems not to be as critical as in previous culture systems. Nevertheless, differences in between batches are still apparent, and we strongly recommend some tests to be performed for obtaining the proper FBS batch. We also strongly advice that a few vials be stored at –80°C, so as to generate long-term stores.

c. In our experience, PS and HEPES can be stored at 4°C with no loss.

d. β-Mercaptoethanol is stable for at least 2 yr when stored as a 14 *M* solution in the dark. The working 10^{-1} *M* solutions will be stored at 4°C under aluminium foil and will be freshly prepared every week.

e. Dexamethasone, TGF-α, and TGF-β will be stored as an aliquot at –20°C. Do not refreeze a thawed aliquot, but store it at 4°C for no longer than 2 wk. Avoid preparing large amounts of TGF-α and TGF-β solutions, as those factors will be much more stable as powder.

2. At least two types of selection can be envisioned:

a. A G418 selection when the retrovirus expresses a NeoR gene. In that case, apply the selection by adding 30 µL of a 100 mg/mL G418 solution/mL of culture medium (the final concentration of G418 being 3 mg/mL) for at least 4 d. In the case where an excessive death occurs, the living cells can be separated from the dead cells and debris through a centrifugation1050*g* for 15 min, no brake) on top of a Lymphocyte Separation Medium (LSM)(Eurobio; density = 1.077 g/mL). The interface will contain mostly living cells.

b. A selection procedure relying on the factor independence that is induced by some transforming oncogenes like the v-*erbB* oncogene *(53)*. In that case, 48 h after infection, simply grow the cells in a LM1 medium devoid of dexamethasone, TGF-α, and TGF-β for a few days. No normal cells will grow in those conditions.

3. The viral suspensions stability should be a permanent concern. Even at –80°C, the viral titer drops with time. Fresh stock solutions of the most used virus should be prepared by a new round of infection every year. In order to minimize the impact of mutations arising through this reiterative infection protocol, generate new viral batches through transfection with the original plasmids every 5 yr or so.

Acknowledgments

The work in our laboratory is supported by the Ligue contre le cancer (Comité départemental du Rhône), the CNRS, the UCBL, the Région Rhône-Alpes (Programme Emergence), the Fondation de France, the Association pour la recherche contre le cancer (ARC), and a joint INSERM/AFM project. I thank Sébastien Dazy for critical reading of the manuscript and Mirielle Deguillien and Faouzi Baklouti (CGMC, UCBL, Villeurbanne) for their invaluable help in setting up the T2ECs transfection assay.

References

1. Gandrillon, O., Rascle, A., and Samarut, J. (1995) The v-*erbA* oncogene: a superb tool for dissecting the involvement of nuclear hormone receptors in differentiation and neoplasia. *Int. J. Oncol.* **6,** 215–231.

2. Sap, J., Munoz, A., Damm, K., et al. (1986) The c-erbA protein is a high-affinity receptor for thyroid hormone. *Nature* **324,** 635–640.

3. Weinberger, C., Thompson, C. C., Ong, E. S., Lebo, R., Gruol, D. J., and Evans, R. M. (1986). The c-erb-A gene encodes a thyroid hormone receptor. *Nature* **324,** 641–646.

4. Rascle, A., Gandrillon, O., Cabello, G., and Samarut, J. (1997) The v-erbA oncogene, in *Oncogenes as Transcriptional Regulators* (Ghysdael, J. and Yaniv, M., eds.) Birkhäuser Publishing Ltd, Basel, pp. 119–165.

5. Mangelsdorf, D. J., Umesono, K., and Evans, R. M. (1994) The retinoid receptors. In *The Retinoids: Biology, Chemistry and Medicine*, (Sporn, M. B., Roberts, A. B., and Goodman, D. S., eds.) Raven Press Ltd, New York, pp. 319–349.

6. Damm, K., Thompson, C. C., and Evans, R. M. (1989) Protein encoded by v-erbA functions as a thyroid-hormone receptor antagonist. *Nature* **339,** 593–597.

7. Sharif, M. and Privalsky, M. L. (1991) v-erbA oncogene function in neoplasia correlates with its ability to repress retinoic acid receptor action. *Cell* **66,** 885–893.

8. Urnov, F. D., Yee, J., Sachs, L., et al. (2000). Targeting of N-CoR and histone deacetylase 3 by the oncoprotein v-ErbA yields a chromatin infrastructure-dependent transcriptional repression pathway. *Embo J.* **19,** 4074–4090.

9. Thormeyer, D. and Baniahmad, A. (1999) The v-erbA oncogene (review). *Int. J. Mol. Med.* **4,** 351–358.

10. Stunnenberg, H. G., Garcia-Jimenez, C., and Betz, J. L. (1999). Leukemia: the sophisticated subversion of hematopoiesis by nuclear receptor oncoproteins. *Biochim. Biophys. Acta* **1423,** F15–F33.

11. Chen, J. D. and Evans, R. M. (1995) A transcriptional co-repressor that interacts with nuclear hormone receptors. *Nature* **377,** 454–457.

12. Hörlein, A. J., Näär, A. M., Heinzel, T., et al. (1995) Ligand-independent repression by the thyroid horone receptor mediated by a nuclear hormone co-repressor. *Nature* **377,** 397–404.

13. Desbois, C., Aubert, D., Legrand, C., Pain, B., and Samarut, J. (1991) A novel mechanism of action for the v-erbA oncogene: abrogation of the inactivation of AP-1 transcription factor by retinoic acid receptor and thyroid hormone receptor. *Cell* **67,** 731–740.

14. Schule, R., Rangarajan, P., Yang, N., et al. (1991) Retinoic acid is a negative regulator of AP-1-responsive genes. *Proc. Natl. Acad. Sci. USA* **88,** 6092–6096.

15. Zhang, X. K., Wills, K. N., Husmann, M., Hermann, T., and Pfahl, M. (1991) Novel pathway for thyroid hormone receptor action through interaction with jun and fos oncogene activities. *Mol. Cell. Biol.* **11,** 6016–6025.

16. Schroeder, C., Gibson, L., Zenke, M., and Beug, H. (1992b) Modulation of normal erythroid differentiation by the endogenous thyroid hormone and retinoic acid receptors: a possible target for v-erbA oncogene action. *Oncogene* **7,** 217–227.

17. Gandrillon, O., Ferrand, N., Michaille, J.-J., Roze, L., Zile, M. H., and Samarut, J. (1994) c-erbAalpha/T3R and RARs control commitment of hematopoietic self-renewing progenitor cells to apoptosis or differentiation and are antagonized by the v-erbA oncogene. *Oncogene* **9,** 749–758.

18. Bauer, A., Mikulits, W., Lagger, G., Stengl, G., Brosch, G., and Beug, H. (1998) The thyroid hormone receptor functions as a ligand-operated developmental switch between proliferation and differentiation of erythroid progenitors. *Embo J.* **17,** 4291–4303.

19. Boucher, P., Koning, A., and Privalski, M. L. (1988) The avian erythroblastosis virus erb A oncogene encodes a DNA—binding protein exhibiting distinct nuclear and cytoplasmic subcellular localizations. *J. Virol.* **62,** 534–544.

20. Mikulits, W., Schranzhofer, M., Bauer, A., et al. (1999) Impaired ferritin mRNA translation in primary erythroid progenitors: shift to iron-dependent regulation by the v-erbA oncoprotein. *Blood* **94,** 4321–4332.

21. Rascle, A., Ghysdael, J., and Samarut, J. (1994) c-erbA, but not v-erbA, competes with a putative erythroid repressor for binding to the CAII promoter. *Oncogene* **9,** 2853–2867.

22. Barettino, D., Bugge, T. H., Bartunek, P., et al. (1993) Unliganded T3R, but not its oncogenic variant, v-erbA, suppresses RAR-dependent transactivation by titrating out RXR. *EMBO J.* **12,** 1343–1354.

23. Chen, H. W., Smitmcbride, Z., Lewis, S., Sharif, M., and Privalsky, M. L. (1993) Nuclear hormone receptors involved in neoplasia: erb-A exhibits a novel DNA sequence specificity determined by amino acids outside of the zinc-finger domain. *Mol. Cell Biol.* **13,** 2366–2376.

24. Subauste, J. S. and Koenig, R. J. (1995) Comparison of the DNA binding specificity and function of v-ErbA and thyroid hormone receptor alpha1. *J. Biol. Chem.* **270,** 7957–7962.

25. Judelson, C. and Privalsky, M. L. (1996) DNA recognition by normal and oncogenic thyroid hormone receptors—unexpected diversity in half-site specificity controlled by non-zinc-finger determinants. *J. Biol. Chem.* **271,** 10,800–10,805.

26. Gandrillon, O., Jurdic, P., Benchaibi, M., Xiao, J.-H., Ghysdael, J., and Samarut, J. (1987) Expression of the v-erb A oncogene in chicken embryo fibroblasts stimulates their proliferation in vitro and enhances tumor growth in vivo. *Cell* **49,** 687–697.

27. Garrido, C., Li, R. P., Samarut, J., Gospodarowicz, D., and Saule, S. (1993). v-erbA cooperates with bFGF in neuroretina cell transformation. *Virology* **192,** 578–586.

28. Iglesias, T., Llanos, S., López-Barahona, M., et al. (1995) Induction of platelet-derived growth factor B/c-sis by the v-erbA oncogene in glial cells. *Oncogene* **10,** 1103–1110.

29. Llanos, S., Iglesias, T., Riese, H. H., Garrido, T., Caelles, C., and Munoz, A. (1996) V-erbA oncogene induces invasiveness and anchorage-independent growth in cultured glial cells by mechanisms involving platelet-derived growth factor. *Cell Growth Differ.* **7,** 373–382.

30. Munoz, A., Wrighton, C., Seliger, B., Bernal, J., and Beug, H. (1993). Thyroid hormone receptor/c-erbA: control of commitment and differentiation in the neuronal/chromaffin progenitor line PC12. *J. Cell Biol.* **121,** 423–438.

31. Cassar-Malek, I., Marchal, S., Altabef, M., Wrutniak, C., Samarut, J., and Cabello, G. (1994) Stimulation of quail myoblast terminal differentiation by the v-erbA oncogene. *Oncogene* **9,** 2197–2206.

32. Kahn, P., Frykberg, L., Brady, C., et al. (1986). v-erb A cooperates with sarcoma oncogenes in leukemic cell transformation. *Cell* **45**, 349–356.
33. Bachnou, N., Laudet, V., Jaffredo, T., Quatannens, B., Saule, S., and Dieterlen-Lievre, F. (1991) Cooperative effect of v-myc and v-erbA in the chick embryo. *Oncogene* **6**, 1041–1047.
34. Metz, T. and Graf, T. (1992) The nuclear oncogenes v-erbA and v-ets cooperate in the induction of avian erythroleukemia. *Oncogene* **7**, 597–605.
35. Khazaie, K., Panayotou, G., Aguzzi, A., Samarut, J., Gazzolo, L., and Jurdic, P. (1991) EGF promotes in vivo tumorigenic growth of primary chicken embryo fibroblasts expressing v-myc and enhances in vitro transformation by the v-erbA oncogene. *Oncogene* **6**, 21–28.
36. Barlow, C., Meister, B., Lardelli, M., Lendahl, U., and Vennström, B. (1994) Thyroid abnormalities and hepatocellular carcinoma in mice transgenic for v-erbA. *Embo J.* **13**, 4241–4250.
37. Meyer, S., LaBudda, K., McGlade, J., and Hayman, M. J. (1994) Analysis of the role of the Shc and Grb2 proteins in signal transduction by the v-ErbB protein. *Mol. Cell. Biol.* **14**, 3253–3262.
38. Disela, C., Glineur, C., Bugge, T., et al. (1991) v-erbA overexpression is required to extinguish c-erbA function in erythroid cell differentiation and regulation of the erb A target CAII. *Genes Dev.* **5**, 2033–2047.
39. Pain, B., Melet, F., Jurdic, P., and Samarut, J. (1990) The carbonic anhydrase II gene, a gene regulated by thyroid hormone and erythropoietin, is repressed by the v-erbA oncogene in erythrocytic cells. *New Biol.* **2**, 284–294.
40. Zenke, M., Munoz, A., Sap, J., Vennstrom, B., and Beug, H. (1990) v-erbA oncogene activation entails the loss of hormone-dependent regulator activity of c-erbA. *Cell* **61**, 1035–1049.
41. Gandrillon, O. and Samarut, J. (1998) Role of the different RAR isoforms in controlling the erythrocytic differentiation sequence. Interference with the v-erbA and p135gap-myb-ets nuclear oncogenes. *Oncogene* **16**, 563–574.
42. Jepsen, K., Hermanson, O., Onami, T. M., et al. (2000). Combinatorial roles of the nuclear receptor corepressor in transcription and development. *Cell* **102**, 753–763.
43. Hayman, M. J., Kitchener, G., Knight, J., McMahon, J., Watson, R., and Beug, H. (1986) Analysis of the autophosphorylation activity of transformation defective mutants of avian erythroblastosis virus. *Virology* **150**, 270–275.
44. Gandrillon, O., Schmidt, U., Beug, H., and Samarut, J. (1999) TGFb cooperates with TGFa to induce the self-renewal of normal erythrocytic progenitors: Evidence for an autocrine mechanism. *EMBO J.* **18**, 2764–2781.
45. Apfel, C., Bauer, F., Crettaz, M., et al. (1992) A retinoic acid receptor alpha antagonist selectively counteracts retinoic acid effects.*Proc. Natl. Acad. Sci. USA* **89**, 7129–7133.
46. Benchaibi, M., Mallet, F., Thoraval, P., et al. (1989) Avian retroviral vectors derived from avian defective leukemia virus: role of the translational context of the inserted gene on efficiency of the vectors. *Virology* **169**, 15–26.

47. Hughes, S. H., Greenhouse, J. J., Petropoulos, C. J., and Sutrave, P. (1987). Adaptor plasmids simplify the insertion of foreign DNA into helper-independent retroviral vectors. *J. Virol.* **61,** 3004–3012.

48. Jurdic, P., Treilleux, I., Vandel, L., et al. (1995) Tumor induction by v-Jun homodimers in chickens. *Oncogene* **11,** 1699–1709.

49. Kawai, S. and Nishizawa, M. (1984) New procedure for DNA transfection with polycation and dimethyl sulfoxide. *Mol. Cell Biol.* **4,** 1172–1174.

50. Vogt, P. K. (1969) Focus assay of Rous sarcoma virus, in *Fundamental Techniques in Virology* (Habel, K. and Salzman, N. P., eds.) Academic Press, New York, pp. 198–211.

51. Moscovici, C., Moscovici, M. G., Jimenez, H., Lai, M. M. C., Hayman, M. J., and Vogt, P. K. (1977) Continuous tissue culture cell lines derived from chemically induced tumors of Japanese quail. *Cell* **11,** 95–103.

52. Samarut, J. (1978) Isolation of an erythropoietic stimulating factor from the serum of anemic chicks. *Exp. Cell Res.* **115,** 123–126.

53. Yamamoto, T., Hihara, H., Nishida, T., Kawai, S., and Toyoshima, K. (1983) A new avian erythroblastosis virus, AEV-H, carries erbB gene responsible for the induction of both erythroblastosis and sarcomas. *Cell* **34,** 225–232.

7

Gene Regulation by Thyroid Hormone Receptors

Helmut Dotzlaw and Aria Baniahmad

1. Introduction

The transfection and overexpression of the cDNA encoding the thyroid hormone receptor (TR) in mammalian cells has shed light into several aspects of the function and biological characteristics of the TR in cells. Using this method, thyroid hormone and TR-responsive genes and response elements were identified. As well, the diversity of TR binding sequences and the effect of heterodimerization with retinoid x receptor (RXR) were elucidated. Furthermore, the functional domains of the TR, its activation domain, hormone-dependent transactivation function, nuclear localization, dimerization properties, and silencing domain could be identified and characterized (**Fig. 1**), for (reviews *see* **refs.** *1–3*). The identification and analysis of functional domains of TR was accomplished mostly by generating fusion proteins of TR parts to a heterologous DNA-binding domain (DBD), such as that of the GAL4-DBD. These fusion proteins permitted the determination of amino acids required for a specific receptor function and allowed the dissection and separate characterization of the various functions of TR *(4,5)*.

In addition, these methods have yielded important information regarding the function of naturally-occurring TR mutants derived from patients with the resistance to thyroid hormone syndrome (RTH) *(6,7)*, as well as the function of the v-erbA oncogene product *(8–10)*. Our understanding of how these mutations affect TR function has increased using the DNA transfection methods. Thus, the methodology of DNA transfection into cells plays a very important role in the functional characterization of the TR.

The introduction of TR cDNA into cells is accomplished by cloning the TR cDNA into a plasmid, which contains mammalian enhancer and promoter sequences upstream of the TR cDNA *(11,12)*, and then introducing this DNA

From: *Methods in Molecular Biology, Vol. 202: Thyroid Hormone Receptors: Methods and Protocols*
Edited by: A. Baniahmad © Humana Press Inc., Totowa, NJ

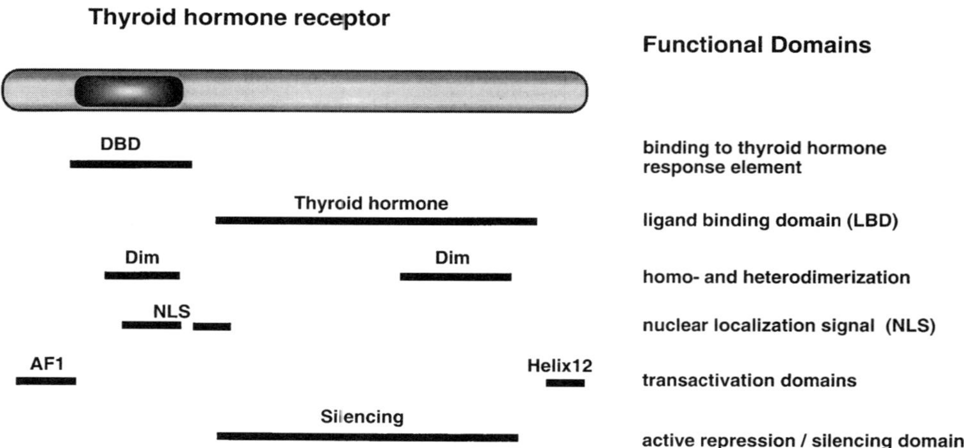

Fig. 1. *Functional domains of the thyroid hormone receptors analyzed by functional assays.* Thyroid hormone receptors encompass a variety of functional domains. The highly conserved zinc finger region is highlighted. The functional DBD extends the zinc finger region and is responsible for recognition and binding to TREs. The ligand-binding domain (LBD) is localized in the receptor carboxy terminus and overlaps to a great extend the silencing domain, which is responsible for active transcription repression in the absence of thyroid hormone. Transactivation domains (AF1 and Helix 12/AF2) are localized in the receptor amino terminus and carboxy terminus, respectively. Nuclear localization is mediated by the nuclear localization signal (NLS). The receptor dimerization for homo- or heterodimers is mediated by the dimerization domains (Dim).

into the cells by one of a variety of methods. TR expression may be analyzed after a few hours up until several days after transfection, this short-term expression is termed "transient expression". The TR cDNA may also be cloned into a plasmid that encodes a selectable marker, usually the cDNA of a gene whose expression confers resistance to a specific antibiotic such as neomycin. If the cells are held under constant selection pressure by growing the cells in the presence of antibiotic, clones that have stably integrated the TR cDNA into the host genome may be selected out and characterized further. This long-term expression is termed "stable expression".

Both transient and stable expression have advantages and disadvantages. Transient expression has the advantages that results are obtained quickly, a selection marker is not required, and because the cDNA is not integrated into the host genome there is no influence on the expression of the cDNA by neighboring genomic DNA sequences. However, depending on the technique used to transfect the cells, many copies of the cDNA may be introduced into the nucleus of each cell. This can make the investigation of the influence of endogenous activator or repressor molecules, which might be expressed at much lower levels than the transfected cDNA, more difficult. Stable expression has the advantages that the expression of the cDNA may be investigated over extended periods of time, and the expression of a single copy of the cDNA per cell may be analyzed. However, stable expression has the disadvantages that the expression of the cDNA may be influenced by neighboring chromatin, a selectable marker is required, clones derived from single cells need to be isolated and analyzed, and typically weeks are required to obtain a result.

Once cells have been transfected, several options are available for the analyses of TR expression and/or function. A direct analysis of the level of expressed TR may be performed by Western blotting of extracts prepared from the transfected cells. DNA/protein interactions, such as the ability of the TR to bind to specific DNA thyroid hormone response elements (TREs), may be investigated through the use of electrophoretic mobility shift assay (EMSA) of such protein extracts. The effect of hormones and/or antihormones on the transcriptional activity of the TR, the determination of minimal promoter DNA sequences required for interaction with the TR, and the effects of overexpressed coactivators–corepressors on TR-mediated transactivation may be studied via cotransfection with reporter genes fused to TREs.

The aim of this chapter is to present various methods for the transfection and analyses of TR action as used in our laboratory.

1.1. Introducing DNA into Mammalian Cells

In general, each cell line has its specific characteristics that does not allow a prediction of which DNA transfer technique might be successfully applied to obtain high levels of expression of the transfected cDNA. Therefore, for each cell line, various DNA-transfer techniques must be tested and optimized. The incubation time of the cells with DNA, treatment of the cells with a "shock" step, such as a glycerol or dimethyl sulfoxide (DMSO) shock, and the time required for optimal expression must be carefully optimized for each cell line. Mostly CV1 cells have been used to express and to analyze the transcriptional properties of the TR, as these cells express no detectable levels of functional endogenous TR-α and TR-β.

We describe several methods of transfecting DNA into cell lines. The $Ca_3(PO_4)_2$ and diethylaminoethyl (DEAE)-dextran methods are inexpensive and applicable to a large number of cell lines ranging from vertebrate to *Drosophila*.

1.1.1. Transfection of Adherent Cells by Calcium Phosphate

The $Ca_3(PO_4)_2$ transfection protocol *(13)* may be used to analyze both transient or stable gene expression. In this protocol, a $Ca_3(PO_4)_2$/DNA precipitate is formed that adheres to the cell surface, and the DNA is taken up into the cell through an as of yet uncharacterized endocytotic process *(14)*. One major advantage of this technique is that the transfected cells contain a representative sampling of the plasmids present in the transfected mixture. This is of particular importance when analyzing the effect on the TR of a second transfected factor such as a coactivator or corepressor. The most critical parameter in this protocol is that the 2X HEPES-buffered saline (HBS) solution is adjusted to the appropriate pH, as the optimum pH range for transfection is very narrow.

1.1.2. Transfection of Adherent Cells with DEAE-Dextran

The DEAE-dextran transfection protocol *(15)* may be used to analyze both transient or stable gene expression. In this protocol, DEAE-dextran/DNA complexes are formed, which adhere to the cell surface, and the DNA is then absorbed through endocytosis. The major advantages of this technique are its simplicity, limited expense, and reproducibility. Critical parameters include the amount of DNA transfected and the ratio of DNA to DEAE-dextran, duration of transfection, and the use of chloroquine. A set of pilot experiments investigating the effects of these variables should be carried out to optimize conditions.

1.1.3. Transfection of Cells in Suspension with DEAE-Dextran

Transfection of cells in suspension with DEAE-dextran offers the same advantages as transfection of adherent cells using this method. The method is particularily suited for the transfection of cells, which grow in suspension, e.g., lymphoid or erythroblast cells, such as HD3 cells.

1.1.4. Transfection of Cells in Suspension by Electroporation

Electroporation *(16)* utilizes high-voltage electric shocks to introduce DNA into cells. The high-voltage electric field generated results in the formation of pores in the cell membrane large enough for macromolecules to pass through into the cell, and ultimately into the nucleus. The method is highly efficient and may be used in both transient and stable transfections. An added advantage is that the copies of DNA introduced per cell may be controlled, which might be important when the expression of only a single DNA copy per cell is desired. However, the major drawback of the method is that larger amount of both cells and DNA is required in transient transfection experiments.

1.1.5. Removal of Endogenous Hormones from Serum by Charcoal Treatment

Transfection analyses often involves determining what effects hormones have on the transcriptional activity of the receptor. For this purpose, the serum is treated with activated charcoal in order to remove endogenous hormones normally present in fetal calf serum.

1.2. Verifying Transfected Protein Expression by Western Blotting

Western blotting involves separating the proteins in a cell extract on the basis of size on a polyacrylamide gel *(17)*, transferring these proteins onto a blotting membrane *(18,19)*, and then probing this blot using an antibody specific to the transfected protein *(21)*. Western blotting provides a means of estimating the expression level of the transfected protein, and whether the expressed protein is of the expected size. It is most useful in such cases where, for example, a functional wild-type protein is compared with a nonfunctional mutant in a reporter assay, to verify that the functional wild-type and nonfunctional mutant proteins are expressed at similar levels in the respective transfected cells.

1.3. EMSA for the Analyses of Protein/DNA Interaction

The EMSA *(22,23)* is used to analyze protein/DNA interactions (**Fig. 2**). It is based on the observation that when protein is bound to DNA, the mobility of the DNA/protein complex is retarded when compared to the mobility of the free DNA probe. Typically, extracts are prepared from cells transfected with cDNA encoding the DNA-binding protein. This extract is then mixed with a labeled oligonucleotide, and the mixture is run out on a nondenaturing polyacrylamide gel. As a control, extracts are prepared from cells not transfected with the DNA-binding protein, and a sample is prepared that contains no added proteins. Upon autoradiography of the gel, samples containing DNA-binding protein will present bands with significantly reduced mobility than those seen in the control lanes.

To obtain large amounts of TR, cells should be used in which the transfected plasmid containing the TR cDNA is replicated. COS1 cells are a derivative of CVI cells harboring the large T-antigen. Plasmids which contain the simian virus 40 (SV40) origin of replication are able to replicate in these cells. Also HEK 293 cells, which are transformed by adenoviral DNA, allow high expression levels. Another source of TR, TR-variants or TR deletions, is through an in vitro transcription–translation system. Typically, we use the TNT-System from Promega, because in this system, TR cDNAs are efficiently transcribed and translated providing that either a T3, T7, or SP6 RNA polymerase promoter and Kozak sequence are present on the plasmid. It should be noted that for most of the TR binding sites, the heterodimer partner of the TR, RXR, must be present. When extracts prepared from cells are used, the amount of RXR in the extract is usually sufficient to interact with TR and to cause a mobility shift. However, when in vitro translated TR is used, in should be pre-incubated with in vitro translated RXR prior to use in the EMSA.

1.4. Reporter Systems and Assays for Reporter Enzyme Activity

Reporter systems are based on the fusion of the promoter sequence of interest (e.g., a TRE) with a gene encoding a reporter molecule. The measurement

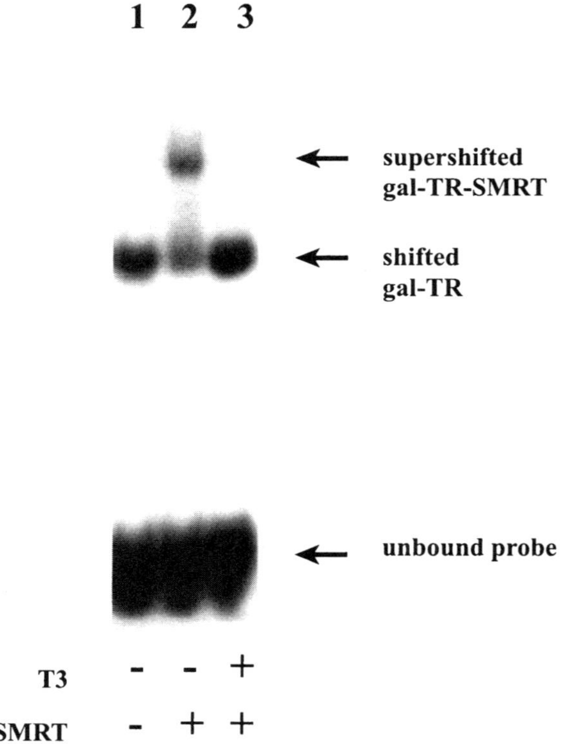

Fig. 2. *EMSA demonstrating the interaction of TR with the corepressor SMRT.* COS cells were transfected with the cDNAs encoding the Gal-TR and/or SMRT and treated with or without thyroid hormone (T3). Cell extracts from transfected CV1 cells were prepared and subjected to EMSA using a labeled UAS, the Gal4 binding site. Lane 2 contains a "supershifted" band of the SMRT/Gal-TR/UAS complex, demonstrating the hormone-sensitive interaction of SMRT with DNA-bound TR *(27)*.

of the level of expression of the reporter molecule synthesized reflects the ability of the promoter sequence of interest to promote transcription. Ideally, the reporter molecule should be easily assayed and should have minimal or no physiological effect on the transfected cell. For normalization, identical amounts of cell extracts *(24)* or ideally an unresponsive reporter is added as a control. The four reporter systems used in our laboratory for measuring TRE transcriptional activity are presented below.

1.4.1. Chloramphenicolacetyltransferase

Chloramphenicolacetyltransferase (CAT) is one of the most widely used and accepted reporter systems. The enzyme catalyzes the transfer of acetyl groups

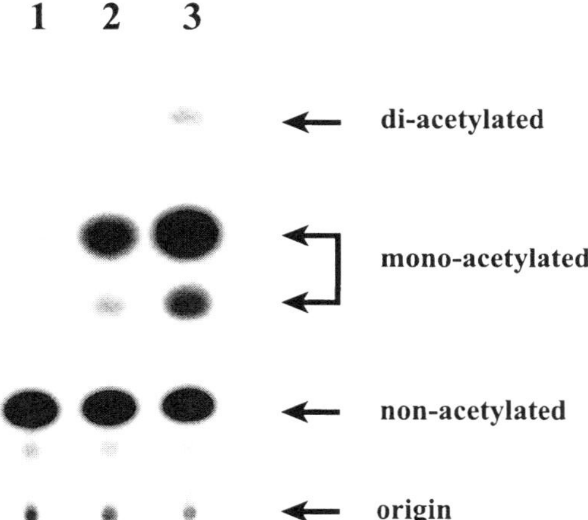

Fig. 3. *CAT Reporter Assay demonstrating the various forms of acetylated chloramphenicol.* Hela cells were transfected with the cDNAs encoding TR and TR mutants in the presence of thyroid hormone together with a reporter construct containing TREs upstream of the CAT gene, TRE3x-tkCAT. Shown are the small spot of loading origin, the non acetylated, the two mono-acetylated, and the one di-acetylated forms of chloramphenicol separated by thin-layer chromatography. Increased expression of CAT leads to an increase in enzyme activity and to acetylation of chloramphenicol. Lane 1 shows low CAT expression and lane 3 shows high CAT expression.

from acetyl coenzyme A to chloramphenicol, and the acetylated form of ^{14}C chloramphenicol is separated from the unacetylated form by thin-layer chromatography (**Fig. 3**)*(25)*. The assay has the advantage that it is reliable and easy, although time-consuming. The disadvantage to the CAT reporter molecule is that the assay is relatively insensitive compared to nonisotopic chemiluminescent reporter systems. The CAT enzyme is very stable, and cell extracts may be heated to 60°C for 10 min prior to use in order to inactivate endogenous enzymes, which utilize acetyl coenzyme A.

1.4.2. Firefly-Luciferase

The reaction catalyzed by firefly luciferase requires luciferin, ATP, Mg^{2+}, and molecular O_2. Light is emitted by an activated luciferin intermediate, with the amount of light output being proportional to the amount of luciferase in the assay *(26)*. The assay is fast, easy, and nonradioactive and is more sensitive than the CAT assay, but has the disadvantage that it requires a relatively expensive piece of equipment, a luminometer, to perform.

Two luciferase assays are detailed in this chapter, each with its own lysis and assay solutions. The Flash Assay is suitable for use in luminometers, which utilize an injector for the addition of substrate to sample, as the kinetics of this reaction consist of a short (few seconds) flash of light, which rapidly decays over the course of a few seconds. The Sustained Assay is suitable for use in both luminometers with or without injectors. In this reaction, the light output is sustained over a period of up to several minutes. The Sustained Assay also has the advantage that it is at least 10-fold more sensitive than the Flash Assay, however, it is more expensive to perform because it contains coenzyme A.

1.4.3. β-Galactosidase

β-Galactosidase catalyzes the hydrolysis of various β-galactosides. The colorimetric assay is based on the hydrolysis of o-nitrophenol-β-D-galactopyranoside (ONPG) to o-nitrophenol and galactose. The absorbance produced by o-nitrophenol, a yellow color, is measured in a spectrophotometer. Although the colorimetric assay is not very sensitive, β-galactosidase is often cotransfected with other plasmids and used to normalize for transfection efficiency between replicate transfections.

1.4.4. Secreted Alkaline Phosphatase

Secreted alkaline phosphatase (SEAP) is a truncated form of the human placental alkaline phosphatase lacking a membrane anchoring domain. The enzyme is secreted from the cell, allowing analysis using a small amount of the cell culture medium. This offers the major advantage that the cells remain intact for further analyses. Also, because SEAP is extremely heat stable, endogenous phosphatase activity can be eliminated by a simple heat treatment of the sample at 65°C. The chemiluminescent assay for SEAP is very sensitive, dephosphorylation of the chemiluminescent phosphatase substrate (CSPD) results in a glow that is measured in a luminometer and remains constant for up to 60 min. We use a commercial assay kit to measure SEAP activity, the Roche SEAP Reporter Gene Assay.

2. Materials

2.1. Methods of Introducing DNA into Mammalian Cells

2.1.1. Transfection of Adherent Cells by Calcium Phosphate

1. 0.25 M CaCl$_2$: autoclave or filter-sterilize.
2. 2X HBS: 280 mM NaCl, 1.5 mM Na$_2$HPO$_4$, 50 mM HEPES. Adjust to pH 7.08 and filter-sterilize.
3. Glycerol mixture: 50 mL 2X HBS, 30 mL glycerol, 120 mL H$_2$O. Autoclave.

2.1.2. Transfection of Adherent Cells with DEAE-Dextran

1. Chloroquine: 100 mM in H$_2$O. Filter-sterilize. Store in 1-mL aliquots at –20°C.

2. DEAE-dextran: 1 mg/mL (Pharmacia) in Tris-buffered saline (TBS).
4. 20X TBS: 0.5 M Tris, 2.74 M NaCl, 100 mM KCl, 14 mM CaCl$_2$, 10 mM MgCl$_2$. Adjust to pH 7.4 and autoclave. When diluting to 1X TBS add 1.2 mL 0.5 M Na$_2$HPO$_4$ per liter 1X TBS, then autoclave.

2.1.3. Transfection of Cells in Suspension with DEAE-Dextran

1. DEAE-Dextran: 1 mg/mL (Pharmacia) in TBS.
2. 20X TBS: 0.5 M Tris, 2.74 M NaCl, 100 mM KCl, 14 mM CaCl$_2$, 10 mM MgCl$_2$. Adjust to pH 7.4 and autoclave. When diluting to 1X TBS add 1.2 mL 0.5 M Na$_2$HPO$_4$ per liter 1X TBS, then autoclave.

2.1.4. Transfection of Cells in Suspension by Electroporation

1. 10X Phosphate-buffered saline (PBS): 1.2 M NaCl, 280 mM Na$_2$HPO$_4$, 25 mM KH$_2$PO$_4$, pH 7.3.

2.1.5. Removal of Endogenous Hormones from Serum by Charcoal Treatment

1. Activated charcoal, Sigma (cat. no. C-9157).
2. Thyroid hormone (T3): a stock solution is dissolved to a concentration of 10^{-2} M in 10 mM NaOH and diluted with sterile water to 10^{-4} M to 10^{-5} M. Add to media at a 1000-fold dilution.

2.2. Verifying Transfected Protein Expression by Western Blotting

1. NETN: 20 mM Tris-HCl, pH 8.0, 150 mM NaCl, 1 mM EDTA, 0.5% Nonidet P40 (NP-40), 10% glycerol.
2. Acrylamide: *bis*-acrylamide solution: 29.2% acrylamide, 0.8% *bis*-acrylamide.
3. 1.5 M Tris-HCl, pH 8.8.
4. 0.5 M Tris-HCl, pH 6.8.
5. 10% sodium dodecyl sulfate (SDS).
6. 5X Tris-glycine running buffer: 15 g Tris, 72 g glycine, 5 g SDS per L (DO NOT adjust pH with acid or base).
7. 5X SDS Sample reducing buffer: 1 mL 0.5 M Tris-HCl, pH 6.8, 1.6 mL glycerol, 1.6 mL 10% SDS, 0.4 mL β-Mercaptoethanol, 2 mL 0.5% bromophenol blue in water, 1.4 mL water.
8. Transfer buffer I: 36.6 g Tris, 100 mL methanol per liter.
9. Transfer buffer II: 3.03 g Tris, 100 mL methanol per liter.
10. Transfer buffer III: 5.2 g 6-amino-n-caproic acid, 100 mL methanol per liter.
11. 10X PBS: 1.2 M NaCl, 280 mM Na$_2$HPO$_4$, 25 mM KH$_2$PO$_4$.
12. PBST: 100 mL 10X PBS, 10 mL Tween-20 per L.

2.3. EMSA for the Analyses of Protein/DNA Interaction

1. Binding buffer: 20 mM HEPES, pH 7.8, 400 mM NaCl, 10% glycerol, 2 mM dithiothreitol (DTT).
2. 40% Acrylamide:bis-acrylamide: 39.2% acrylamide, 0.8% *bis*-acrylamide.
3. 10X Tris-glycine buffer: 250 mM Tris, 1.92 M glycine.

4. Probe elution buffer: 100 m*M* KCl, 10 m*M* Tris-HCl, pH 7.5.
5. Bio-Rad protein assay: cat. no. 500-0001.
6. 0.25 *M* Tris-HCl, pH 7.8.
7. Calf thymus DNA: 1 mg/mL in water. Sheared and denatured by boiling 10 min and placing on ice.
8. Poly dI*dC: 1 mg/mL (Pharmacia) in water.
9. 4X Loading dye: 20% glycerol, 0.01% bromophenol blue, 0.01% xylene cyanol.

2.4. Reporter Systems and Assays for Reporter Enzyme Activity

2.4.1. Chloramphenicolacetyltransferase (CAT) Assay

1. 10X PBS: 1.2 *M* NaCl, 280 m*M* Na_2HPO_4, 25 m*M* KH_2PO_4, pH 7.3.
2. Acetyl coenzyme A: 10 mg coenzyme A in 3 mL 0.25 *M* Tris-HCl, pH 7.8.
3. 0.25 *M* Tris-HCl, pH 7.8.
4. Ethyl acetate.
5. 93:7 Chloroform:methanol.

2.4.2. Firefly-Luciferase Assay

1. 10X PBS: 1.2 *M* NaCl, 280 m*M* Na_2HPO_4, 25 m*M* KH_2PO_4 pH 7.3.
2. Flash lysis buffer: 25 m*M* Tris-HCl, pH 7.5, 8 m*M* $MgCl_2$, 1 m*M* EDTA, 1% Triton X-100, 15% glycerol, add fresh before use 1 m*M* DTT, 0.2 m*M* phenylmethylsulfonyl fluoride (PMSF).
3. Flash Substrate Solution: 5 mg D-luciferin, 400 μL 100 m*M* ATP, 2.5 mL 1 *M* Tris-HCl, pH 7.5, per 50 mL.
4. Sustained lysis buffer: 50 m*M* Tris-Ac, pH 7.8, 10 m*M* MgAc, 0.1 m*M* EDTA, 1% Triton X-100, 15% glycerol, add fresh before use 4 m*M* DTT, 0.2 m*M* PMSF.
5. Sustained substrate solution: 15 mg coenzyme A, 5 mg D-luciferin, 500 μL 100 m*M* ATP, 2.5 mL 1 *M* Tris-Ac, pH 7.8, per 50 mL.

2.4.3. β-Galactosidase Assay

1. Z-Buffer: 60 m*M* Na_2HPO_4, 40 m*M* NaH_2PO_4, 10 m*M* KCl, 1 m*M* $MgSO_4$, 50 m*M* β-Mercaptoethanol (added fresh before use).
2. ONPG substrate: 4 g ONPG in 1 L 0.1 *M* $NaPO_4$, pH 7.5. Store in aliquots at –20°C.
3. 0.5 *M* Na_2CO_3.

2.4.4. SEAP Assay

As per manufacturer's instructions, Roche Molecular Biochemicals SEAP Reporter Gene Assay, Product Number 1 779 842.

3. Methods

3.1. Methods of Introducing DNA into Mammalian Cells

3.1.1. Transfection of Adherent Cells by Calcium Phosphate

We describe the calcium phosphate method below for a 10-cm culture dish. However, the size of the dish can also be larger or smaller. In these cases, the volumes of solutions and the amount of DNA can be scaled up or down, depending on the surface area of the dish being used. The surface area of a dish is calculated by πr^2, where r represents the radius of the dish.

1. Seed out cells (*see* **Note 1**).
2. For each transfection prepare the following mixture:
     380 µL 0.25 *M* $CaCl_2$.
     20 µL (1–20 µg) DNA in water.
3. Add 400 µL 2X HBS to mixture, drop-wise over a duration of approx 30 s with continuous gentle mixing, at room temperature (*see* **Note 2**).
4. Allow the precipitate to form for 30 min at room temperature.
5. Add precipitate drop-wise to the cells.
6. After 4 h to overnight incubation with the $CaPO_4$/DNA precipitate, a glycerol shock step, which increases transfection efficiency in some cell lines, may be performed.

Glycerol Shock Step

7. Aspirate media.
8. Add 2 mL glycerol mixture per dish, incubate for 3 min at room temperature.
9. Aspirate glycerol mixture and wash each dish twice with 4 mL PBS per wash.
10. Add media and incubate cells for 1–4 d at 37°C in 5% CO_2.

3.1.2. Transfection of Adherent Cells with DEAE-Dextran

1. Seed out cells (*see* **Note 1**).
2. For each dish prepare the following mix:
     50 µL DNA in TE, pH 7.5.
     350 µL TBS.
     450 µL DEAE-dextran (1 mg/mL in TBS).
3. Aspirate medium from cells.
4. Wash with 5 mL TBS, aspirate (*see* **Note 3**).
5. Add DNA/DEAE-dextran to cells evenly.
6. Incubate 30 min at room temperature (*see* **Note 4**).
7. Aspirate mixture.
8. Wash with 4 mL PBS.
9. Add 10 mL medium.
10. Add 10 µL chloroquine if required. (*see* **Note 5**).
11. Incubate 40–72 h at 37°C, 5% CO_2.

3.1.3. Transfection of Cells in Suspension with DEAE-Dextran

1. Grow cells on dishes to late log phase—needed are 5×10^6 cells/transfection.
2. Trypsinize cells and harvest into medium.
3. Centrifuge for 5 min at 200*g*.
4. Resuspend cell pellet in 1/2 original volume TBS.
5. Count cells and distribute 5×10^6 cells/transfection into microfuge tubes.
6. Centrifuge 5 min at 200*g*.

7. Resuspend cells:
   25 µg DNA in 20 µL TE, pH 7.5.
   150 µL TBS.
   220 µL DEAE-dextran.
8. Incubate for 1 h at room temperature, carefully mixing every 10 min.
9. Transfer cell suspension into 10-cm dishes containing 8 mL medium per dish.
10. Harvest and analyze after 40–60 h incubation at 37°C, 5% CO_2.

3.1.4. Transfection of Cells in Suspension by Electroporation

1. Grow cells to late log phase—needed are 5×10^7 cells/transfection.
2. Trypsinize cells, harvest in medium, and centrifuge for 5 min at 200g.
3. Resuspend cells in 1/2 original volume ice-cold PBS.
4. Centrifuge for 5 min at 200g.
5. Resuspend cells to 5×10^7 cells/mL in ice-cold PBS.
6. Transfer 0.5 mL cell suspension into desired number of cuvets, place on ice.
7. Add DNA (10–40 µg for transient transfection).
8. Mix by flicking cuvet and place on ice.
9. Place cuvette in electroporator and shock with setting capacitor 25 µF and 1200 V for 0.4-cm cuvets.
10. Place cuvette on ice for 10 min.
11. Dilute transfected cells 20-fold in medium, rinsing cuvette to obtain all cells.
12. Plate cells, harvest, and analyze after 50–60 h incubation at 37°C, 5% CO_2.

3.1.5. Removal of Endogenous Hormones from Serum by Charcoal Treatment

1. Heat-inactivate the serum by heating at 56°C for 30 min.
2. Add 25 g of activated charcoal per 500 mL serum.
3. Stir slurry for 2 h at room temperature.
4. Centrifuge twice at 7500g for 15 min each.
5. Filter serum through a 0.8-µm filter (prefilter), aliquot into 50-mL aliquots.
6. Filter through a 0.2-µm (sterile) filter into medium prior to use.

3.2. Verifying Transfected Protein Expression by Western Blotting

3.2.1. Preparation of Protein Extracts using NETN

1. Wash dish twice with 5 mL PBS per wash.
2. Add 1 mL ice-cold NETN.
3. Scrape cells into microfuge tube.
4. Perform 3 freeze/thaw cycles in liquid nitrogen/37°C water bath.
5. Centrifuge for 5 min a 14,000 rpm in a microfuge centrifuge.
6. Transfer supernatant to a fresh tube.
7. Determine protein concentration using the Bradford Protein Assay (*see* **Subheading 3.2.2.**).

Table 1
Preparation of Gel Mixtures

	Separating			Stacking
Gel percentage	12%	10%	7.5%	4%
Acrylamide:bis solution	4.0 mL	3.3 mL	2.5 mL	1.3 mL
H_2O	3.4 mL	4.1 mL	4.9 mL	6.1 mL
1.5 *M* Tris-HCl, pH 8.8	2.5 mL	2.5 mL	2.5 mL	—
0.5 *M* Tris-HCl, pH 6.8	—	—	—	2.5 mL
10% SDS	100 µL	100 µL	100 µL	100 µL
40% Ammonium persulfate	50 µL	50 µL	50 µL	50 µL
N,N,N',N',-Tetramethyl-ethylenediamine (TEMED)	5 µL	5 µL	5 µL	10 µL

3.2.2. Bradford Protein Determination

This assay is based on the observation that the absorbance maximum of the dye Coomassie Blue shifts from 495 to 595 nm upon binding to protein *(24)*.

1. Prepare samples:
   10 µL Whole cell extract supernatant.
   790 µL 0.25 *M* Tris-HCl, pH 7.8.
   200 µL 5X Concentrated Coomassie Blue dye solution.
2. Mix and allow to stand for for 15 min.
3. Measure absorbance at 595 nm relative to Blank (no added extract). A protein standard curve should be prepared using various amounts of bovine serum albumin (BSA), from 1–50 µg. (*see* **Note 6**).
4. Calculations: µg protein in sample = $A_{595} \times 40$.
   Concentration of protein = µg protein in sample/µL sample added to assay.

3.2.3. Denaturing Polyacrylamide Gel

1. Assemble the gel apparatus and prepare the separating gel mixture (*see* **Note 7**). The volumes shown in **Table 1** correspond to those required for a 1-mm-thick minigel, Bio-Rad, or equivalent.
2. Pour the separating gel and overlay with ethanol.
3. Allow to polymerize for 30 min.
4. Drain ethanol and allow to dry for a few minutes.
5. Prepare the stacking gel mixture, pour gel and insert comb.
6. Allow to polymerize for 15 min.
7. Assemble apparatus and fill with 1X Tris-glycine running buffer.
8. Prepare samples by diluting desired amount to 20 µL in water, then adding 5 µL 5X SDS reducing buffer.

9. Boil samples for min, centrifuge briefly at 14,000 rpm in a microfuge centrifuge, and load onto gel.
10. Run gels at 150 V for 1 h (until bromophenol blue dye front reaches the bottom of the gel).
11. Proceed with the Semi-Dry Blotting Procedure (*see* **Subheading 3.2.4.**).

3.2.4. Semi-Dry Blotting Procedure

1. Cut 15 sheets of filter paper to the dimensions of the gel.
2. Pour a small amount of Transfer buffers I-III into trays.
3. Wet 6 sheets of filter paper in Transfer buffer I.
4. Wet 3 sheets of filter paper in Transfer buffer II.
5. Wet 6 sheets of filter paper in Transfer buffer III.
6. Wet polyvinylidene fluoride (PVDF)-membrane (Millipore Immobilon-P) by immersing in methanol for 10 s, then place into Transfer buffer II.
7. Wet gel in Transfer buffer II.
8. Wet the anode with Transfer buffer I and assemble the blot on the Anode in the following order:
   6 Sheets of filter paper in Transfer buffer I.
   3 Sheets of filter paper in Transfer buffer II.
   Membrane.
   Gel.
   6 Sheets of filter paper in Transfer buffer III.
9. Wet the cathode with Transfer buffer III.
10. Assemble the blotting apparatus and blot at 1 mA/cm^2 of gel for 1 h.

3.2.5. Immune Detection

1. Disassemble the blotting apparatus and place the membrane into PBS with Tween 20 (PBST) containing 5% skim milk (from nonfat dried milk powder, e.g., Carnation).
2. Incubate for 1 h at room temperature, or overnight at 4°C.
3. Wash membrane 2× briefly with PBST.
4. Incubate membrane with primary antibody for 1 h at room temperature in PBST. (*see* **Note 8**).
5. Wash blot 3× 5 min per wash in PBST.
6. Incubate blot in appropriately conjugated secondary antibody for 1 h at room temperature.
7. Wash at least 5× 5 min in PBST and procede with chemiluminescent detection *(21)* following the manufacturer's instructions (e.g., Amersham ECL Kit).

3.3. EMSA for the Analyses of Protein/DNA Interaction

3.3.1. Whole Cell Extract

1. Aspirate medium from cells.
2. Add 1 mL ice-cold PBS to each dish and scrape cells into a microfuge tube.
3. Centrifuge for 1 min at 2000 rpm at 4°C in a microfuge centrifuge.

4. Resuspend cell pellet in 200 µL binding buffer by repeated pipeting.
5. Freeze in liquid nitrogen and allow to thaw on ice.
6. Centrifuge for 30 min at 4°C at 14,000 rpm in a microfuge centrifuge.
7. Transfer supernatant into a fresh microfuge tube.
8. Determine protein concentration using the Bradford Protein Determination (*see* **Subheading 3.2.2.**).

3.3.2. Labeling of Probe

1. Prepare the following mixture:
   1 µL (1–3 pmol) Double-stranded oligonocleotides (*see* **Note 10**).
   5 µLg^{32}P ATP.
   1 µL 10X T4 PNK Buffer (as supplied with enzyme).
   2 µL H$_2$O.
   1 µL (10 U) T4 PNK.
2. Incubate for 30 min at room temperature.
3. Separate from free ATP through a spin-column or by running a native polyacrylamide gel, cutting out the DNA probe, and eluting with 300–500 µL of elution buffer.

3.3.3. Native Polyacrylamide Gel

1. Prepare gel mixture:
   10 mL 40% Acrylamide solution.
   6 mL 10X Tris-glycine buffer.
   3.6 mL Glycerol.
   40.4 mL H$_2$O.
   220 µL 40% Ammonium persulfate.
   35 µL TEMED.
2. Pour into pre-assembled gel chamber and insert comb.
3. Cool 1X Tris-glycine buffer and gel chamber in cold room prior to run.

3.3.4. Binding of Sample to Prob

1. Prepare the following mixture:
   1 µL poly dI*dC (1 mg/mL, Pharmacia).
   0.25 µL 1 mg/mL calf thymus DNA.
   8 µL H$_2$O.
   5 µL Whole cell extract (4 mg, dilute with binding buffer if necessary).
   1 µL Labeled DNA probe (10 fmol/µL).
2. Incubate for 15 min on ice.
3. Assemble gel chamber.
4. Prerun gel at 100 V for 30 min. We routinely run EMSA gels in the cold room.
5. Add 5 µL 4X loading dye in a separate slot to monitor migration in the gel (*see* **Note 11**).
6. Load samples and run gel at 100–200 V until the bromophenol blue dye front reaches the bottom of the gel.
7. Dry gel, cover with plastic wrap, and expose to X-ray film.

3.4. Reporter Systems and Assays for Reporter Enzyme Activity

3.4.1. CAT Assay

1. Aspirate medium.
2. Wash dishes twice with 5 mL PBS per wash.
3. Add 300 µL 0.25 *M* Tris-HCl, pH 7.8.
4. Scrape cells into a microfuge tube.
5. Perform 3 freeze/thaw cycles in liquid nitrogen/37°C water bath.
6. Centrifuge for 5 min and transfer supernatant to a fresh microfuge tube (*see* **Note 12**).
7. Make desired amount of sample to 150 µL with 0.25 *M* Tris-HCl, pH 7.8. (*see* **Note 13**).
8. Add 20 µL acetyl-coenzyme A plus 0.5 µL ^{14}C-chloramphenicol (Amersham).
9. Incubate several h up to overnight at 37°C.
10. Add 0.5 mL ethylacetate, vortex mix vigorously, and centrifuge for 1 min.
11. Transfer upper phase to a fresh microfuge tube.
12. Allow to evaporate to dryness in a Speed-Vac.
13. Add 10 µL ethylacetate and spot onto TLC plate (Polygram Sil G, Macherey-Nagel).
14. Chromatograph in 93:7 chloroform:methanol until the front is 3/4 way to the top.
15. Dry plate, wrap in plastic-wrap, and expose to X-ray film for few hours up to overnight.

3.4.2. Firefly-Luciferase Assay

1. Aspirate medium.
2. Wash dishes twice with 5 mL PBS per wash.
3. Add 400 µL lysis buffer per 10-cm dish.
4. Incubate for 10 min at room temperature.
5. Scrape cells and transfer to a microfuge tube (*see* **Note 14**).
6. Centrifuge for 10 min at 4°C at 14,000 rpm.
7. Transfer supernatant to a fresh microfuge tube (*see* **Note 15**).
8. Measure the amount of light emitted by 100 µL sample in a luminometer after injection of 100 µL luciferin substrate.

3.4.3. β-Galactosidase Assay

1. Aspirate medium.
2. Wash dishes twice with 5 mL PBS per wash.
3. Add 1 mL TEN and incubate 10 min at room temperature.
4. Scrape cells and transfer to a microfuge tube.
5. Centrifuge for 5 min at 2000 rpm in a microfuge centrifuge and discard supernatant.
6. Add 350 µL Z-Buffer to pellet.
7. Perform 3 freeze/thaw cycles in liquid nitrogen/37°C water bath.
8. Centrifuge for 10 min at 4°C and transfer supernatant to a fresh microfuge tube.
9. Dilute desired amount of sample to be measured to 1 mL with Z-Buffer. (*see* **Note 16**).

10. Incubate 5 min at 37°C.
11. Add 200 µL ONPG solution.
12. Prepare blank which contains 1 mL Z-Buffer plus 200 µL ONPG.
13. Incubate at 37°C until visibly yellow.
14. Stop the β-galactosidase reaction by adding 500 µL 0.5 M Na_2CO_3.
15. Measure absorbance at 420 nm against blank.

3.4.4. SEAP Assay

1. Transfer medium from transfected cells into a microfuge tube. As a negative control use medium from untransfected cells.
2. Centrifuge medium for 2 min (*see* **Note 17**).
3. Dilute the cleared medium 1:40 in dilution buffer (2.5 µL medium plus 97.5 µL dilution buffer).
4. Incubate for 30 min at 65°C.
5. Centrifuge for 1 min at 14,000 rpm.
6. Add 100 µL inactivation buffer.
7. Incubate for 5 min at room temperature.
8. Add 50 µL substrate solution (425 µL Solution II plus 25 µL CSPD).
9. Incubate for 10 min at room temperature
10. Measure in luminometer.

4. Notes

1. In general, cells should be seeded out prior to transfection at a density that allows their growth for a further 3 to 4 d. Cells that are very confluent will not take up DNA as efficiently as cells growing in the logarithmic phase, and cells may detach from the dish if too confluent. For CV1 cells, we generally use 5×10^5 cells per 10-cm dish.
2. To reproducibly control the rate at which the 2X HBS is added to each sample, we use a peristaltic pump, which is controlled with a timer to drip the 2X HBS into the samples *(14)*.
3. It is important at this step to aspirate the TBS completely, yet being careful to avoid drying-out of the cells.
4. Gently mix the DNA solution every 5–10 min during the incubation.
5. If chloroquine is used, aspirate media after 5–6 h, wash dishes with 4 mL PBS each dish, and add 10 mL medium.
6. Samples yielding an absorbance at 595 nm of greater than 0.8 should be remeasured using a smaller amount of sample.
7. Decide what Separating Gel percentage is required. Proteins >150 kDa are best resolved on a 7.5% gel, proteins <50 kDa on a 12% gel, and proteins between 50 kDa and 150 kDa on a 10% gel.
8. The optimal concentration of primary antibody to use must be determined empirically.
9. Denature probe by heating at 65°C for 5 min and then allowing to anneal by cooling slowly to 4°C.

10. Addition of unlabeled single-stranded DNA avoids the binding of single strand binding proteins to the probe. This is important when small oligonucleotides are used as a probe, which will denature easily to single strands, especially when palindromic sequences are used.
11. Do not add loading dye to the samples. The dye changes the salt concentration and osmolarity of the binding-reaction. There is sufficient glycerol in the binding reaction to facilitate loading into the well.
12. Extacts may be stored at –20°C at this stage.
13. Samples may be heated at 60°C for 10 min and centrifuged for 5 min at 14,000 rpm to minimize endogenous activity.
14. Some protocols call for a freeze/thaw step at this stage, however, in our lysis buffer this is not necessary.
15. Extracts can be stored at –80°C at this stage.
16. The extracts prepared for the CAT or luciferase assays can be used directly by addition of Z-Buffer and ONPG.
17. The medium may be stored at –20°C at this stage.

References

1. Baniahmad, A., Burris, T. P., and Tsai, M.-J. (1994) *The Nuclear Hormone Receptor Superfamily* (Tsai, m.-J. and O´Malley, B. W., eds.) Landes Company, Austin, TX., CRC Press, pp. 1–24.
2. Parker, M. G. and White, R. (1996) Nuclear receptors spring into action. *Nat. Struct. Biol.* **3,**113–115.
3. Schwabe, W. R. (1996) Transcriptional control: How nuclear receptors get turned on. *Curr. Biol.* **6,** 372–374.
4. Baniahmad, A., Köhne, A. C., and Renkawitz, R. 1992. A transferable silencing domain is present in the thyroid hormone receptor, in the v-erbA oncogene product and in the retinoic acid receptor. *EMBO J.* **11,** 1015–1023.
5. Baniahmad, A., Leng, X., Burris, T. P., Tsai, S. Y., Tsai, M.-J. und O'Malley, B.W. (1995) The τ4 activation domain of the thyroid hormone receptor is required for release of a corepressor(s) necessary for transcriptional silencing. *Mol. Cell Biol.* **15,** 76–86
6. Sakurai, A., Miyamoto, T., Refetoff, S., and DeGroot, L. J. (1990) Dominant negative transcriptional regulation by a mutant thyroid hormone receptor-beta in a family with generalized resistance to thyroid hormone. *Mol. Endocrinol.* **4,** 1988–1994.
7. Chatterjee, V. K., Nagaya, T., Madison, L. D., Datta, S., Rentoumis, A., and Jameson, J. L. 1991 Thyroid hormone resistance syndrome. Inhibition of normal receptor function by mutant thyroid hormone receptors. *J. Clin. Invest.* **87,** 1977–1984.
8. Baniahmad, A., Tsai, S. Y., O'Malley, B. W., and Tsai, M. J. (1992) Kindred S thyroid hormone receptor is an active and constitutive silencer and a repressor for thyroid hormone and retinoic acid responses. *Proc. Natl. Acad. Sci. USA* **89,** 10,633–10,637.

9. Collingwood, T. N., Adams, M., Tone, Y., and Chatterjee, V. K. (1994) Spectrum of transcriptional, dimerization, and dominant negative properties of twenty different mutant thyroid hormone beta-receptors in thyroid hormone resistance syndrome. *Mol. Endocrinol.* **8,** 1262–1277.

10. Yen, P. M. and Chin, W. W. (1994) Molecular mechanisms of dominant negative activity by nuclear hormone receptors. *Mol. Endocrinol.* **8,** 1450–1454.

11. Sap J., Munoz A., Damm K., Goldberg Y., Ghysdael J., Leutz A., Beug H., Vennstrom B. 1986 The c-erb-A protein is a high-affinity receptor for thyroid hormone. *Nature* **324,** 635–640.

12. Weinberger, C., Thompson, C. C., Ong, E. S., Lebo, R., Gruol, D. J., and Evans, R. M. 1986 The c-erb-A gene encodes a thyroid hormone receptor. *Nature* **324,** 641–646.

13. Wigler, M., Pellicer, A., Silverstein, S. and Axel, R. 1978. Biochemical transfer of single-copy eucaryotic genes using total cellular DNA as donor. *Cell* **19,** 725–731.

14. Steiner, C. and Kaltschmidt, C. (1989) An automated method for calcium-mediated gene transfer. *Trends Gen.* **5,** 138.

15. Lopata, M. A., Cleveland, D. W., and Sollner-Webb, B. (1984) High-level expression of a chloramphenicol acetyltransferase gene by DEAE-dextran-mediated DNA transfection coupled with a dimethylsulfoxide or glycerol shock treatment. *Nucl. Acids Res.* **12,** 5707.

16. Wong, T. K. and Neumann, E. (1982) Electric field mediated gene transfer. *Biochem. Biophys. Res. Commun.* **107,** 584–587.

17. Laemmli, U. K. (1970) Cleavage of structural proteins during the assembly of the head of bacteriophage T4. *Nature* **227,** 680–685.

18. Towbin, H., Staehelin, T. and Gordon, J. (1979) Electrophoretic transfer of proteins from polyacrylamide gels to nitrocellulose sheets: Procedures and some applications. *Proc. Natl. Acad. Sci. USA* **76,** 4350–4354.

19. Bjerrum, O. J. and Shafer-Nielson, C. (1986) Buffer systems and transfer parameters for semidry electroblotting with a horizontal apparatus, *Electrophoresis '86* (Dunn, M. J., ed.) VCH Publishers, Deerfield Beach, FL. pp. 315–327.

20. Burnette, W. N. (1981) Western blotting: Electrophoretic transfer of proteins from sodiumdodecylsulfate polyacrylamide gels to unmodified nitrocellulose and radiographic detection with antibody and radioiodinated protein A. *Anal. Biochem.* **112,** 195–203.

21. Schneppenheim, R., Budde, U. Dahlmann, N., and Rautenberg, P. (1991) Luminography - a new, highly sensitive visualization method for electrophoresis. *Electrophoresis* **12,** 367–372.

22. Garner, M. M. and Revzin, A. (1981) A gel electrophoresis method for quantifying the binding of proteins to specific DNA regions: Application to components of the Escherichia coli lactose operon system. *Nucl. Acids Res.* **9,** 3047–3060.

23. Fried, M. and Crothers, D. M. (1981) Equilibria and kinetics of lac repressor-operator interactions by polyacrylamide gel electrophoresis. *Nucl. Acids Res.* **9,** 6505–6525.

24. Bradford, M. M. (1976) A rapid and sensitive method for the quantification of microgram quantities of protein utilizing the principle of protein dye binding. *Anal. Biochem.* **72,** 248–254.

25. Gorman, C. M., Moffat, L. F., and Howard, B. H. (1982) Recombinant genomes which express chloramphenicol acetyltransferase in mammalian cells. *Mol. Cell Biol.* **2,** 1044–1051.
26. Gould, S. J. and Subramani, S. (1988) Firefly luciferase as a tool for molecular and cell biology. *Anal. Biochem.* **7,** 5–13.
27. Baniahmad, A., Dressel, U., and Renkawitz, R. (1998) Cell specific inhibition of RARα silencing by the AF2/τc activation domain can be overcome by the co-repressor SMRT, but not N-CoR. *Mol. Endocrinol.* **12,** 504ENDASH512

8

Molecular Analysis of Human Resistance to Thyroid Hormone Syndrome

Sunnie M. Yoh and Martin L. Privalsky

1. Introduction

Resistance to thyroid hormone (RTH) syndrome is an inherited human endocrine disease, which is manifested as a failure to respond properly to elevated circulating thyroid hormone *(1–4)*. RTH syndrome behaves as an autosomal dominant trait, and has been mapped at the molecular level to a diverse array of mutations within the thyroid hormone receptor (TR)-β locus *(1–4)*. As detailed elsewhere in this review series, TRs bind to specific DNA sequences, denoted thyroid hormone response elements (TREs) and regulate transcription of adjacent target genes in response to thyroid hormone *(5–7)*. TRs typically repress transcription in the absence of hormone and activate transcription in the presence of hormone *(8–10)*. These bipolar transcriptional properties of the receptor are mediated by the receptor's ability to recruit ancillary polypeptides, denoted corepressors and coactivators, to the target promoter *(11–16)*. Corepressors and coactivators modulate transcription both by covalent modification of the chromatin template and by direct interactions with components of the general transcriptional machinery. Corepressors are typically recruited by TRs in the absence of hormone, whereas binding of hormone to the TRs results in a release of the corepressors and a recruitment of the coactivator polypeptides *(11–16)*.

Virtually all of the TR mutations associated with RTH syndrome result in the synthesis of aberrant TRβ receptors that retain the ability to repress, but that are impaired in the ability to activate transcription in response to hormone *(1–4)*. In fact, many of the RTH-TRβ mutant receptors function as dominant-negatives, interfering *in trans* with transcriptional activation by normal TRs; this dominant-negative interference is believed to play an important role in the pathophysiology of RTH syndrome *(17–22)*. The molecular basis of the dominant-negative pheno-

From: *Methods in Molecular Biology, Vol. 202: Thyroid Hormone Receptors: Methods and Protocols*
Edited by: A. Baniahmad © Humana Press Inc., Totowa, NJ

type, in many cases, appears to reflect an inability of the RTH-TRβ mutants to release from corepressor (and to acquire coactivator) in response to physiological levels of thyroid hormone *(21,23–29)*. For the majority of the RTH-TRβ mutants studied to date, the failure to release corepressor and to recruit coactivator reflects a reduced or abolished affinity of the mutant receptor for thyroid hormone. However, for at least some alleles associated with RTH syndrome, the affinity of the mutant receptor for hormone is near-normal, but instead the triggering mechanisms that couple hormone binding to corepressor release (and to coactivator acquisition) are disrupted by the genetic lesion *(21,23–29)*.

It should be noted that RTH syndrome in humans is a complex trait. Several distinct forms of disease can be distinguished clinically, such as pituitary-specific vs generalized RTH syndrome *(1–4)*. Otherwise, identical genetic lesions in TRβ can result in different disease phenotypes in different patients, suggesting that genetic polymorphisms outside of the TRβ loci, or epigenetic effects, may also play a role in modulating the penetration, severity, and clinical presentation of the disease *(30–32)*. The TRβ locus is also expressed as variety of different receptor isoforms due to alternative mRNA splicing, and the effects of a given RTH syndrome mutation may differ when examined in the different isoform backgrounds (e.g. *[33]*). Finally, it must be noted that the different mutations, which are associated with RTH syndrome, can result in alterations to DNA binding, receptor dimerization, and receptor stability *(1–4)*. We will describe how RTH-TRβ genes are isolated from patients as molecular clones, how the mutations are characterized, how the affinity of wild-type and mutant TRs for hormone are characterized experimentally, and how the interactions of wild-type and mutant TRs with one another and with auxiliary proteins, such as corepressors and coactivators, are analyzed.

2. Materials

2.1. Molecular Cloning of TR Sequences from RTH Syndrome Patients

1. Genomic DNAs extracted from RTH patients' blood.
2. 10X Polymerase chain reaction (PCR) Buffer: 100 mM Tris-HCl, pH 8.3, 500 mM KCl, 0.01% gelatin.
3. 50 mM MgCl$_2$.
4. dNTP mixture: a mixture of 2 mM each of dATP, dGTP, dCTP and dTTP
5. *Taq* DNA polymerase.
6. Streptavidin-coated magnetic beads (DYNABEAD M-280 streptavidin: DYNAL, Wirral, UK), and/or Quik change site-directed mutagenesis kit (Stratagene, La Jolla, CA).
7. Primers:
   a. Oligonucleotide primers flanking exons 4–10 of human TRβ for the primer sequences, *see* **ref.** *34*). So as to facilitate later sequencing, one oligonucle-

otide of each primer set can be biotinylated at its 5' end, which can be subsequently captured by streptavidin-coated magnetic beads after amplification to generate a single-stranded DNA template.

   b. Mutagenic primers containing appropriate mutations (used in PCR-based and in Quik change site-directed mutagenesis schemes.

8. Standard reagents for recombinant DNA research, including plasmid vectors, restriction endonucleases and buffers, and agarose gel electrophoresis reagents.

2.2. Measurement of the Hormone Affinity of Mutant and Wild-Type TRs

2.2.1. In Vitro Transcription–Translation System

1. TnT Reticulolysate System (Promega, Madison, WI).
2. pSG5-TRβ wild-type and pSG5-RTH-TRβ constructs.
3. ^{35}S-methionine (Trans-label, 1000 Ci/mmol; NEN, Boston, MA).

2.2.2. Filter Binding Assay

1. L-3, 5, 3'-[^{125}I]-Triiodothyronine (3000 mCi/mg; NEN, Boston, MA).
2. L-3,5,3'-Triiodothyronine (Sigma, St. Louis, MO) prepared as an 1 mM stock in 100% ethanol.
3. Millipore HAWPO-2500 filters (Millipore Corp, Bedford, MA).
4. T3 Binding buffer: 20 mM Tris-HCl, pH 7.5, 50 mM NaCl, 2 mM EDTA, 5 mM 2-Mercaptoethanol.
5. 10% Nonidet P-40 (NP40).
6. Vacuum manifold filtration device.
7. Gamma counter.

2.2.3. Protease Resistance Assay

1. Trypsin and elastase (recrystallized; Roche, Indianapolis, IN).
2. 50 mM Tris-HCl, pH 7.5.
3. Reagents and apparatus for sodium dodecyl sulfate-polyacrylamide gel electrophoresis (SDS-PAGE).
4. Equipment for gel autoradiography or phosphorimager analysis.

2.3. Analysis of Homo- and Heterodimer Formation by RTH-TRs

2.3.1. Preparation of Sf9 Cell Nuclear Extracts:

1. Baculovirus stocks expressing wild-type and RTH-TRβs.
2. Sf9 cells.
3. Ex-cell 401 medium (JRH Biosciences, Lenex, KS) supplemented with 10% fetal bovine serum (FBS) (Life Sciences Technology, Grand Island, NY).
4. 1X Tris-buffered saline (TBS): 10 mM Tris-HCl, pH 7.5, 100 mM NaCl.
5. 10% NP-40.
6. Sf9 Lysis buffer A: 10 mM HEPES, pH 7.9, 10 mM KCl, 0.1 mM EDTA, pH 8.0, 0.1 mM EGTA, 1 mM dithiothreitol, (DTT), 0.5 mM phenylmethane-sulfonyl fluoride (PMSF).

7. Sf9 Lysis Buffer B: 20 mM HEPES, pH 7.9, 400 mM KCl, 1 mM EDTA, 1 mM EGTA, 1 mM DTT, 1 mM PMSF, 10% glycerol.

2.3.2. Radiolabeling of DNA Probes

1. Annealed oligonucleotides containing TRE sequences.
2. [α-^{32}P]dGTP (3000 Ci/mmol; NEN, Boston, MA).
3. Klenow fragment of DNA polymerase I (New England Biolabs, Beverly, MA).
4. 10X Klenow buffer: 100 mM Tris-HCl, pH 7.5, 50 mM MgCl$_2$, 75 mM DTT.
5. A mixture of dATP, dTTP, and dCTP, 5 mM each.
6. Sephacryl S-200HR (Sigma, St. Louis, MO).
7. Scintillation counter.

2.3.3. Electrophoretic Mobility Shift Assay (EMSA)

1. 10X Tris-borate EDTA (TBE): 890 mM Tris, 890 mM boric acid, 20 mM EDTA, pH 8.0.
2. EMSA Binding buffer: 12 mM Tris-HCl, pH 7.5, 3.5% glycerol, 77 mM KCl, 2.3 mM MgCl$_2$, 15.4 µg/µL bovine serum albumin (BSA), 0.15 µg/µL poly-deoxyinosine-deoxycytosine (poly-dI-dC).
3. Polyacrylamide gel electrophoresis apparatus.
4. Gel dryer apparatus.
5. Facilities for autoradiography or phosphorimager analysis.

2.4. Analysis of the Cofactor Interaction

2.4.1. Expression of Glutathione-S-Transferase (GST)-Fusion Proteins

1. Appropriate pGEX-KG constructs containing corepressor or coactivator domains.
2. BL-21 Competent *Escherichia coli.*
3. 100 mM Sterile isopropyl-β-D-thiogalactopyranoside (IPTG).
4. Branson Sonifer 250 with microtip or comparable apparatus.

2.4.2. GST-Binding Reactions

1. ^{35}S-labeled in vitro translated RTH-TRs (*see* **Subheading 2.2.**).
2. Phosphate-buffered saline with Triton (PBST) Buffer: 150 mM NaCl, 16 mM Na$_2$HPO$_4$, 4 mM NaH$_2$PO$_4$, 2 mM EDTA, 1% Triton X-100, 1 mM DTT.
3. HEMG buffer: 40 mM HEPES, pH 7.5, 100 mM KCl, 0.2 mM EDTA, 5 mM MgCl$_2$, 0.1% NP-40, 10% glycerol, 1.5 mM DTT.
4. Glutathione-coupled agarose (Sigma, St. Louis, MO).
5. 10 mM Glutathione in 50 mM Tris-HCl, pH 8.0.
6. 100 mg/mL bovine serum albumin (BSA).
7. Complete protease inhibitor cocktail (Roche, IN).

2.4.3. Mammalian 2-Hybrid Interactions

1. pSG5GAL4-DBD and pSG5GAL4-AD constructs containing appropriate regions of corepressors, coactivators, and RTH-TRβs.

2. pCMV-lacZ plasmid.
3. pGL-GAL4-17mer luciferase reporter.
4. pSG5 empty vector.
5. CV-1 cells.
6. Dulbeco's-modified Eagles' medium (DMEM) (Life Sciences Technology, Grand Island, NY) supplemented with 0.1% bicarbonate and 10% fetal bovine serum (FBS).
7. Lipofectin (Life Sciences Technology, Grand Island, NY).
8. 5X Lysis buffer (Promega, Madison, WI).
9. Luciferase assay kit (Promega, Madison, WI).
10. β-Galactosidase activity assay reagent: 0.5 mM chlorophenol red-β-D-galacto-pyranoside (CRPG), 0.06 M Na$_2$HPO$_4$, 0.06 M NaH$_2$PO$_4$, 0.01 M KCl, 1 mM MgCl$_2$, 20 mM 2-Mercaptoethanol.
11. TD 20/20 luminometer (Turner Designs, Sunnyvale, CA) or equivalent.

2.4.4. Biotin-DNA Protein Complex Assay

1. Biotin-conjugated oligonucleotides containing suitable TREs.
2. Nuclear extracts of Sf9 cells expressing the receptors of interest (*see* **Subheading 2.3.**)
3. ^{35}S-labeled in vitro synthesized cofactors (*see* **Subheading 2.2.**), streptavidin-conjugated agarose (Pierce, Rockford, IL).
4. PBST buffer.
5. HEMG buffer.
6. 100 mg/mL BSA.
7. Protease inhibitors.
8. SDS-PAGE reagents and apparatus as detailed in **Subheading 2.2.**

3. Methods

3.1. Molecular Cloning of TR Sequences from RTH Syndrome Patients

RTH syndrome patients are identified initially in the clinic as individuals manifesting aberrantly high levels of circulating T3 and T4 thyronine hormone, normal or elevated levels of thyroid-stimulating hormone (TSH), and overt or compensated hypothyroidism *(1–4)*. Typically, once patients are diagnosed with RTH syndrome, genomic DNA is isolated from blood samples and subjected to PCR (PCR) so as to recover adequate material for molecular cloning and for sequence analysis (e.g. *[34]*). These approaches generally focus on exons 4 to 10 of the *TRβ* gene, which encode all but the first 8 amino acids of the open reading frame of the receptor. The sequences of the patient *TRβ* genes are then compared to the wild-type TRβ sequence to identify any specific base pair substitutions, deletions, or additions (*see* **Note 1**). Once a specific TRβ mutation has been identified, it can be introduced into a variety of expression vectors for further experimental characterizations. Usually, the RTH syndrome TRβ mutations, identified from genomic subclones, are introduced into the full-length receptor open reading frame by a two-step PCR or similar site-directed mutagenesis procedure. This process of abstracting the individual mutations and placing them in an otherwise wild-type TRβ background is important for distinguishing mutations that are a cause of the syndrome from genetic polymorphisms in the human population that have no detectable effect on receptor function.

3.1.1. Isolation of RTH-TRβ DNA and Sequence Analysis

3.1.1.1. EXTRACT LEUKOCYTE DNA FROM A BLOOD SAMPLE OF THE RTH PATIENT

A variety of protocols or commercially available kits can be utilized for this purpose, such as the SNAP Whole Blood DNA isolation kit (Invitrogen, Carlsbad, CA), the QIAamp DNA Blood kits (Qiagen, Valencia, CA), or the Dynabeads DNA Direct kit (DYNAL, Wirral, UK).

3.1.1.2. DESIGN SUITABLE OLIGONUCLEOTIDE PRIMER PAIRS

18 to 24 base long suitable oligonucleotide primer pairs are designed representing sequences flanking individual exons within the wild-type *TRβ* gene. A biotinylated oligonucleotide can be substituted for one of each primer pair to simplify the subsequent DNA sequence analysis (*see* below).

3.1.1.3. TYPICAL PCRS ARE PERFORMED AS FOLLOWS *(34)*

1. Mix 1 μg of genomic DNA and 10 pmol of each oligonucleotide primer in a 50-μL reaction volume containing 5 μL of 10X PCR buffer, 2 mM MgCl$_2$, 0.5 mM dNTPs, and 0.5 U of *Taq* DNA polymerase(*see* **Note 2**).
2. Initially denature the DNA samples at 94°C for 3 min.
3. Follow by 30 cycles of (denaturation at 94°C for 30 s, primer annealing at 60°C for 30 s, and primer extension at 72°C for 30 s).
4. Complete the final PCR cycle by a final extension at 72°C for 15 mins.

3.1.1.4. PCR PRODUCTS SEQUENCED DIRECTLY

1. Capture the biotinylated PCR products using streptavidin-coated magnetic beads.
2. Denature the DNA with NaOH to release the nonbiotinylated strand *(35)*.
3. The immobilized biotinylated strand can then be sequenced using an internal oligonucleotide primer and standard dideoxy-sequencing methodologies.

3.1.1.5. PCR PRODUCTS DIRECTLY CLONED INTO PLASMID VECTOR

Alternatively, the PCR products can be cloned directly into a suitable plasmid vector, using TA Cloning (Invitrogen, Carsbad, CA), PCR-Script (Stratagene, La Jolla, CA), Sure-clone (Amersham Pharmacia, Littlechalfront, UK), or similar approaches as recommended by the manufacturers.

1. Recover the resulting recombinant DNA clones by transformation of competent *E. coli.*
2. Identify recombinant clones by minipreparation.

3. Restriction digest analysis.
4. The recombinant clones can then be analyzed by standard DNA dideoxy-sequencing methodologies as adapted for double-stranded DNA templates *(36)*.

3.1.2. Introduction of RTH Mutations into a Wild-Type TRβ Background

3.1.2.1. PCR-BASED SITE-DIRECTED MUTAGENESIS *(37)*

1. Create a set of four oligonucleotide primers for the mutagenesis reactions as follows: design two, partially complementary "mutagenic" primers that overlap the region of the TRβ DNA sequence to be altered and that incorporate the RTH genetic lesion (**Fig. 1**, primers b and c). When introducing a single base substitution into the DNA, the mutagenic primers should each be at least 24 bases long; longer primers, 32 bases or greater in length, should be used if 2 or 3 contiguous base substitutions are to be introduced into the DNA simultaneously, or if a deletion or insertional mutation is to be introduced. Each mutagenic primer should contain at least 15 bases complementary to TRβ sequences downstream (i.e., 3') of, and at least 8 bases complementary to TRβ sequences upstream (i.e., 5') of the site of the desired genetic lesion. Design two additional, "flanking" primers that bracket the entire region to be amplified (**Fig. 1**, primers a and d). The flanking primers, typically 24–32 nucleotides long, should be designed to be fully complementary to the template sequence and adjusted in length and base composition so as to have an annealing temperature comparable to that of the mutagenic primers.
2. Generate of intermediate products AB and CD (**Fig. 1**). Set up two separate PCRs on ice as in **Table 1**. Incubate the samples 1 min at 95°C to denature the DNA, then subject to 25 cycles of (denaturation at 94°C for 1 min, primer annealing at 50°C for 1 min, and primer extension at 72°C for 1 min), followed by a final 72°C extension for 15 mins. Resolve the PCR products by standard agarose gel electrophoresis and visualize by ethidium bromide staining. Using a clean razor blade or scapel, physically excise the slice of agarose containing the DNA fragments of interest, then extract the DNA from the agarose using a Strataprep (Stratagene, La Jolla, CA), Concert (Gibco-BRL, Grand Island, NY), or similar DNA isolation procedure.
3. Overlap extension reaction. Set up a PCR identical to **step 2**, but use aliquots of both the AB and CD PCR products in place of the original template DNA, and use only the flanking primers a and d (i.e., omit the mutagenic primers b and c). When the two intermediate PCR products are mixed and allowed to anneal, the top strand of AB can pair up with the bottom strand of CD and act as primers on one another to make the full-length product AD (**Fig. 1**). Inclusion of primers a and d allows the amplification of this recombinant AD fragment. After 25 cycles of PCR amplification, the final PCR product should be size-fractionated by agarose gel electrophoresis, excised, and purified by the Strataprep or Concert procedures.
4. Cleave the purified fragment AD, bearing the RTH mutation, with appropriate restriction enzymes and use this DNA fragment to replace the corresponding TRβ wild-type sequences in the pSG5-TRβ clone. We often use the pSG5 vector because it permits expression of the mutant receptors both by in vitro tran-

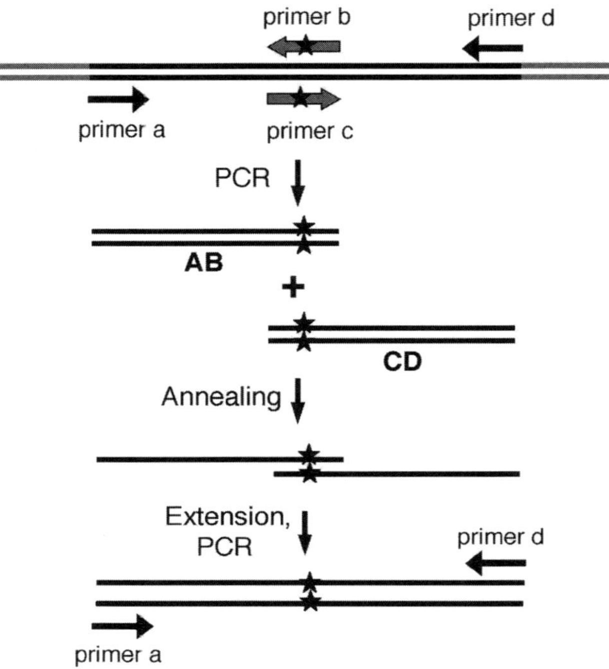

Fig. 1. Site-directed mutagenesis by overlap extension. The template DNA, encoding the wild-type TRβ sequence, is displayed as a double line. The 4 primers described in the text are shown as arrows, from 5' to 3'. The two mutagenic primers (b and c) contain the desired base changes (indicated as stars). The two flanking primers (a and d) are also shown. Two intermediate PCR products, AB and CD, are produced from the template in separate reactions by using a and b primers, and c and d primers, respectively (**Subheading 3.1.2.**). The intermediate PCR products are subsequently purified, annealed to each other, and then amplified by PCR using primers a and d as shown.

Table 1
Set Up for Two Separate PCRs

Ingredients	PCR1 (product AB)	PCR2 (product CD)
10X PCR buffer	5 µL	5 µL
50 mM MgCl$_2$	1.5 µL	1.5 µL
dNTPs (2 mM each)	5 µL	5 µL
5' primer (10 µM)	primer a (5 µL)	primer c (5 µL)
3' primer (10 µM)	primer b (5 µL)	primer d (5 µL)
pSG5-human TRβ DNA	250 ng	250 ng
H$_2$O	to 50 µL	to 50 µL
Taq DNA polymerase	1.5 U	1.5 U

scription–translation, and by transfection of mammalian cells in culture. Alternatively, this same strategy can be used to introduce the mutated sequences directly into a variety of other TRβ propagation or expression vectors. The identity of the final recombinants should be confirmed by DNA sequence analysis; typically over 75% of the recovered recombinants contain the desired mutation.

3.1.2.2. QUIK CHANGE SITE-DIRECTED MUTAGENESIS

An alternative to the 2-step PCR methodology for site-directed mutagenesis is the use of the "Quik-Change" kit methodology (Stratagene). This approach avoids the need for subsequent subcloning by using Pfu DNA polymerase II to replicate both plasmid strands with high fidelity and without displacing the mutagenic oligonucleotide primers. The parental plasmids, which are methylated due to their prior propagation in *dam*(+) *E. coli*, are preferentially digested by incubation with Dpn I endonuclease. The remaining, nicked circular DNAs, enriched for the desired mutation(s), are then transformed directly into competent *E. coli* cells. The procedure generates mutants with greater than 80% efficiency.

3.2. Measurement of the Hormone Affiity of Mutant and Wild-Type TRs

Two different methods to measure the T3 binding affinity of TRs are described here. The first method, referred to as a filter binding assay, measures directly the receptor's ability to bind ^{125}I-radiolabeled T3 hormone (e.g., *[38,39]*). When performed carefully and coupled to a Scatchard analysis, this method can generate a reasonably accurate affinity constant (Ka) of receptor for T3. However, it should be noted that high backgrounds of nonspecific hormone binding, and poor reproducibility, can cause problems when employing the filter-binding technique. The second method of measuring hormone binding affinity exploits the change in conformation that takes place in TRs on hormone binding (e.g., *[40]*). TRs, in common with many other nuclear hormone receptors, undergo significant conformational changes on binding to hormone that result in an overall condensation of the structure of the receptor. This compaction on binding to hormone is manifested as a greatly enhanced resistance to protease digestion compared to that observed in the absence of hormone (**Fig. 2**). Although an indirect assessment of the hormone binding affinity of TRs, the protease resistance assay is relatively simple, reproducible, and sensitive.

3.2.1. Filter Binding Assay

1. The easiest method to produce sufficient RTH-TR protein for analysis is to use a suitable in vitro transcription–translation protocol. For this purpose, the full-length TRβ open reading frame should be subcloned into an vector that contains promoters for appropriate phage RNA polymerases, such as those for the T3, T7,

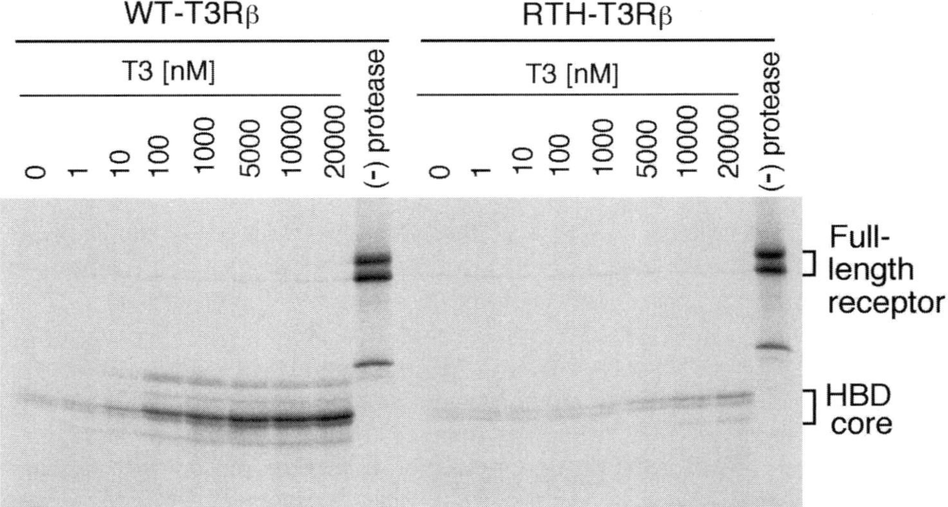

Fig. 2. Hormone-binding assay by the protease resistance method. Wild-type TRβ and a RTH-mutant were synthesized as ^{35}S-radiolabeled proteins by transcription/translation in vitro. Samples of each protein preparation were then adjusted to the T3 hormone concentrations indicated above the panel and were incubated with trypsin as described in **Subheading 3.2.2.** Aliquots of each receptor, maintained in the absence of trypsin, were included for comparison ["(-) protease"]. The incubations were terminated, and the radiolabeled products were resolved by SDS-PAGE and visualized by phosphorimager analysis. The positions of the intact TRs ("Full-length receptor") and of the protease-resistant hormone-binding domain core ("HBD core") are indicated. The full-length TRβ produces two distinct polypeptides when translated in vitro, apparently due to an internal translational start.

or SP6 polymerases; we typically use the pGEM series or pSG5, together with the Promega TnT kit. Use 1 to 2 µg of plasmid DNA in a 50-µL reaction volume, and allow the in vitro transcription and translation to proceed for 90 min at 30°C. Alternative methods, such as expressing the TRβ allele of interest in bacterial cells, in *Saccharomyces cerevisae*, or in mammalian cells, can also be used.

2. Incubate the products of the TnT reactions (0.5–2 µL per assay) for 1 h at 22°C in 10 µL of T3 binding buffer containing from 0.01–10 nM ^{125}I-radiolabeled T3 thyronine (specific activity of 3,000 mCi/mg). Also prepare a parallel set of reactions that include a 1000-fold excess of unlabeled T3 thyronine (for determination of nonspecific binding).

3. For each binding reaction: place a Millipore HAWP02500 filter into a vacuum manifold filtration device and prewet with 3.5 mL of T3 binding buffer. Add the radiolabeled T3 binding reaction and draw into the filter by vacuum suction. Quickly wash the filter with 3 changes of 5 mL each T3 Binding Buffer. Remove

the filter from the vacuum manifold device and count the retained isotope in a gamma counter. The assay should be repeated at least 3 times, and the mean and standard deviation of the triplicate experiments should be determined (*see* **Note 3**).

4. Subtract the nonspecific binding (radiolabel bound to the filter in the presence of a 1000-fold excess of unlabeled hormone) from the specific binding (radiolabel bound to the filter in the absence of cold competitor) for each T3 concentration. Use a Scatchard analysis to generate an apparent K_d for each mutant receptor.
5. The same reactions should also be performed utilizing an unprogrammed reticulocyte lysate as a negative control.

3.2.2. Protease Resistance Assay

It is good idea to subject each receptor to digestion with several proteases possessing different cleavage specificities so as to minimize possible experimental artifacts. The following protocol is a standard protocol for trypsin and elastase but can be adapted to virtually any protease. The concentration of proteases and/or duration of the digestion time can be modified accordingly (*see* **Note 4**).

1. ^{35}S-labeled TRβ proteins are synthesized in vitro using the coupled transcription–translation system described above, and stored on ice.
2. Arrange a series of 6 microfuge tubes containing, for example, 0, 20, 200, 2000, 20,000, or 200,000 mol of nonradioactive T3 thyronine in 15 μL of 50 mM Tris-HCl, pH 7.5, per tube. Use ethanol carrier in place of hormone where appropriate. Chill the tubes on ice and add 1 μL of the in vitro TRβ translation product to each tube. Mix and incubate on ice for 5 min.
3. Add 0.125 μg of protease (trypsin or elastase) in 4 μL of 50 mM Tris-HCl, pH 7.5, per tube. This yields a final T3 hormone concentration of 0, 1, 10, 100, 1000, or 10,000 nM, respectively.
4. Mix. Then sequentially transfer the tubes to a 22°C water bath and incubate each for 10 min. Terminate the reactions sequentially by addition of 20 μL of concentrated SDS-PAGE sample buffer to each tube, followed by rapid freezing on dry ice.
5. Heat the samples for 10 min at 98°C and resolve the protein products by SDS-PAGE (using a 12% polyacrylamide gel). A protease-resistant "core" polypeptide, representing the C-terminal hormone-binding domain of the receptor, should be detectable under high hormone conditions, which is not observed in the absence of hormone (**Fig. 2**, wild-type TRβ). The hormone concentration at which formation of this core protease-resistant polypeptide equals 50% of maximum can be used as a surrogate measure of the receptor's affinity for T3 hormone. In contrast, RTH-TRβ mutants with severely impaired hormone binding affinity will remain highly sensitive to protease at even very high T3 concentrations (**Fig. 2**, RTH-TRβ).

3.3. Analysis of Homo- and Heterodimer formation by RTH-TRs

TRs bind to DNAs either as homodimers or as heterodimers with the retinoid X receptor (RXR) *(5)*. A simple, yet effective, way of analyzing protein–

nucleic acid interactions is the EMSA *(41)*. EMSA is based on the fact that binding of protein to DNA results in a retardation of the electrophoretic mobility of the DNA in a nondenaturing polyacrylamide gel. The assay involves the incubation of a protein of interest with a radiolabeled, linear double-stranded DNA probe, followed by gel electrophoresis to separate protein/DNA complexes from free DNA (**Fig. 3A,B**). Due to the high resolution power of the polyacrylamide gels, protein DNA complexes with different molecular masses can be separated from each other; for example, homodimers of TRβs usually migrate faster in the gel than do heterodimers of TRβ and RXR (**Fig. 3B**). This difference in electrophoretic mobility allows discrimination between receptor homodimers and heterodimers; use of antibodies against TRs or against RXRs in "supershift" experiments can further confirm the identity of the homo- and heterodimer complexes (*see* **Note 5**). TRs and RXRs synthesized by in vitro transcription–translation (**Subheading 3.2.1.**) can be used for EMSA; however, receptor preparations derived from recombinant baculovirus/Sf-9 cell preparations typically produce stronger signals, and lower backgrounds (*see* **Note 6**).

3.3.1. Preparation of Nuclear Extracts of Sf9 Cells Expressing RTH-TRβs

1. Protocols and reagents for the generation of recombinant baculovirus stocks by in vivo recombination can be found in a variety of commercial kits, such as the Baculogold transfection kit (Pharmigen, San Diego, CA). It requires several wk to generate recombinant bacluovirus stocks, each expressing a different RTH-TRβ mutant protein, but once created the viral stocks can be stored and used repeatedly.
2. To prepare recombinant TRβ or RXR proteins, begin by seeding 6×10^6 Sf9 cells in a 100-mm culture plate in Ex-cell 401 media supplemented with 10% FBS. After the cells have attached to the plate, add 0.5–1 mL of high titer viral stock (use a multiplicity of infection of 5 to 10) and incubate the plates for 48 h at 27°C.
3. Harvest by washing the cells off the plate by gently pipeting. Collect the cells by centrifugation at 1500*g* for 5 min at room temperature. Wash the cells twice with 10 mL of TBS buffer. Resuspend the cell pellet in 1 mL of TBS buffer, transfer to an microfuge tube, and collect the cells by centrifugation for 15 s in a microfuge. Chill on ice.
4. Resuspend the cell pellet in 400 μL of ice-cold Sf9 lysis buffer A by gentle pipeting. Allow cells to swell on ice for 15 min, then add 25 μL of 10% NP40 and vigorously vortex mix the tube for 10 s. Centrifuge the lysate in a microfuge for 30 s at 4°C.
5. Carefully remove and discard the supernatant. Resuspend the nuclear pellet in a 50 μL ice-cold Sf9 lysis buffer B. Agitate the tube vigorously at 4°C for 15 min on a shaking platform.
6. Centrifuge the nuclear lysate for 10 min at 12,000*g* in a microfuge at 4°C to pellet debris. Discard the pellet. Snap-freeze and store the supernatant (denoted

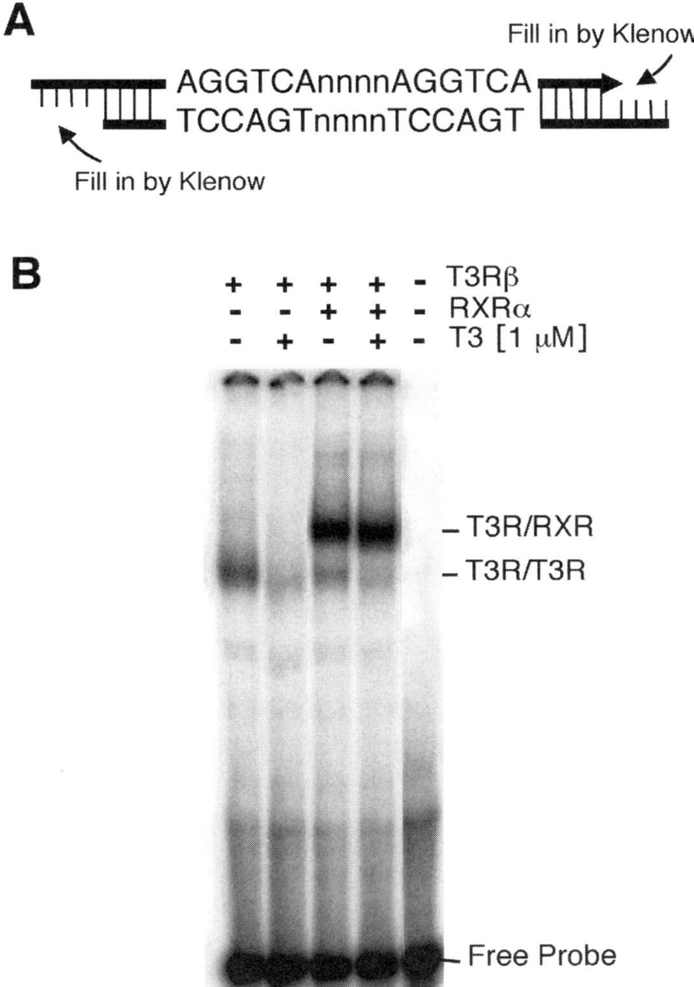

Fig. 3. EMSA. (**A**) Radiolabeling of the DNA probe. A typical DNA probe for use in EMSA analysis is depicted. Two partially complementary oligonucleotides with 5' over-hangs are annealed so as to produce 5' overhangs, as shown. The annealed oligonucle-otides are then labeled with ^{32}P-labeled dNTPs by use of the Klenow fragment of DNA polymerase I (**Subheading 3.3.2.**). The AGGTCAnnnnAGGTCA sequence represents a "direct-repeat 4" TRE that is recognized with high affinity by virtually all TR isoforms. The direction of the arrows indicate the 5' to 3' direction. (**B**) Characterization of DNA binding by wild-type TRβ using EMSA. A radiolabeled DR-4 DNA probe was incubated in the presence of TRβ alone, TRß together with RXRα, or with a control protein prepa-ration lacking receptors, in the presence or absence of T3 hormone as indicated (**Sub-heading 3.3.3.**). The resulting protein/DNA complexes were resolved by native PAGE and visualized by phosphorimager analysis. The positions of free DNA probe, and of the TR homo- and heterodimers complexed with DNA, are indicated to the right of the panel.

"nuclear extract") in 10-µL aliquots at –80°C. Thaw quickly before use and keep on ice. Avoid subjecting nuclear extracts to multiple freeze-thaw cycles.

3.3.2. Radiolabeling the DNA Probes

Prepare by chemical synthesis two partially complementary oligonucleotides, designed to create a suitable double-stranded TRE. These oligonucleotides should be at least 24 bases long, and when annealed, should create 5' overhangs 1–4 bases in length (**Fig. 3A**). We often use oligonucleotides containing TCGA overhangs, which, in addition to permitting their use in EMSA, also allow the introduction of these TREs into reporter vectors containing *Sal*I or *Xho*I sites. The 5' overhangs are filled-in by use of the Klenow fragment of DNA polymerase I and ^{32}P-labeled dNTPs to create a radiolabeled DNA probe. A TRE based on an AGGTCAnnnnAGGTCA sequence (a "direct-repeat 4") is suitable and will bind virtually all TR isoforms with high affinity; however a variety of variations of this sequence can also function as TREs and can be used in these assays *(42)*. Anneal the oligonucleotides by mixing 5 µg each in 20 µL of 40 m*M* Tris-HCl (pH 7.5), 10 m*M* MgCl$_2$, 50 m*M* NaCl. Heat the sample to 75°C for 5 mins, then gradually reduce the temperature to 22°C over a period of 1 h (we use a PCR-type temperature cycler with a programable ramp function for this purpose).

A typical reaction for use in creating a radiolabeled EMSA probe is as follows: 10 µL of 10X Klenow buffer, 10 µL of a mixture of dTTP, dATP, and dCTP (5 m*M* each), 5 µL of [α-^{32}P]dGTP (specific activity of 3000 Ci/mmol), 2 µg of annealed oligonucleotides, 4 U of Klenow Fragment, adjust the final vol to 100 µL with water.

Mix and incubate the reaction at 22°C for 20 min. Add 5 µL of 0.5 *M* EDTA, pH 8.0, and 5 µL of DNA sample dye to terminate the reaction. Separate the labeled DNA probe from free [α-^{32}P]dGTP on a Sepharacyl S-200HR column (2–5 mL bed volume; either gravity or "spin" column methods can be used. Collect the radiolabeled probe eluting in the void volume (approx 200 µL); avoid the fractions containing bromophenol blue dye; these represent the unincorporated isotope. Count a 2-µL aliquot in a scintillation counter to determine the efficiency of the radiolabeling.

3.3.3. EMSA

1. Prepare a 5% nondenaturing polyacrylamide gel (30:1 acrylamide:bis-acrylamide) in 0.25X TBE buffer. We utilize gels that are approx 14 × 14.5 × 0.16 cm (width × height × thickness) with 0.6-cm-wide wells. Place the gel in a vertical electrophoresis apparatus, and prerun at 130 V for 45 min (or until the current drops to 12 to 13 mA) prior to loading the samples (*see* **Note 7**).
2. A range of different TR and RXR concentrations and ratios should be tested. Typically, 1–2 µL of each receptor preparation is aliquoted into prechilled

0.5 mL microfuge tubes, each containing 13 µL of EMSA binding buffer. Receptor preparations, isolated by the recombinant baculovirus Sf9 cell technique, often need to be diluted from 1:20 to 1:100 in Sf9 lysis buffer B prior to use. TRs and RXRs synthesized by in vitro transcription and translation usually do not require dilution.

3. Equilibrate the samples on ice for 10 min. Then add 60,000 to 80,000 counts per min (cpm) of radiolabeled probe in 4 µL of EMSA binding buffer per reaction. Mix, transfer the sample to 22°C to initiate the binding reactions, and incubate for 25 min.

4. Load the binding reactions into the wells of the prerun gel by underlaying the electrode buffer. An aliquot of bromophenol blue (0.01% in EMSA buffer) can also be loaded on an empty lane to monitor the electrophoretic separation (*see* **Note 8**).

5. Perform the electrophoresis at 200 V for 90 min, or until the bromophenol blue tracking dye migrates to a position 2 cm from the bottom of the gel. Remove the gels from the electrophoresis apparatus, and dry on Whatmann 3MM paper using a vacuum gel dryer (e.g., Bio-Rad, Richmond, CA). The dried gels can then be visualized by autoradiography (using intensifying screens) and/or by phosphorimager analysis (**Fig. 3B**).

6. It is important to include several negative controls to eliminate the possibility of nonspecific protein/DNA interactions. These controls should include the use of a radiolabeled DNA probe lacking a functional TRE and the use of protein extracts derived from Sf9 cells infected by a nonrecombinant baculovirus (*see* **Note 9**).

3.4. Analysis of the Cofactor Interaction

Two widely employed methods of studying receptor–cofactor interactions are the GST-pull-down and the mammalian 2-hybrid procedures. The GST-pull-down is an in vitro assay, whereas the mammalian 2-hybrid is performed in vivo. Both methods have advantages and limitations. The GST-pull-down protocol permits study of protein–protein interactions in a relatively pure and well-characterized system. Unfortunately expression of GST-fusion proteins much larger than 80,000 in molecular weight is often problematic due to poor yields and proteolytic degradation in the *E. coli* host. Use of a protease-impaired *E. coli* strain such as BL-21 can help minimize, but cannot fully eliminate this degradation problem. In the case of the mammalian 2-hybrid assay, protein-protein interactions can be characterized in an in vivo context that is likely to more accurately reflect normal physiology. However, the 2-hybrid interaction assay, of necessity, includes all the components of the cell; as a consequence, indirect interactions, as well as direct ones, may score by this procedure (*see* **Note 10**). It should be noted that both the GST-pull-down and 2-hybrid methods analyze protein–protein interactions in the absence of an appropriate DNA binding site. Therefore, we also describe a biotin/DNA pull-down assay that permits the characterization of the receptor–cofactor interaction in the context of DNA binding. In this procedure, the receptor–cofactor complexes are assembled on a biotin-tagged DNA containing a suitable TRE. The protein/DNA complexes are then isolated using streptavidin-conjugated agarose, and the protein components of the complex are resolved by SDS-PAGE and visualized by radiolabeling or immunoblotting. This method also has the advantage of permitting the use of full-length native polypeptides.

3.4.1. GST-Pull-Down Assay

3.4.1.1. SUBCLONING COFACTORS INTO PLASMIDS

Subclone appropriate regions of cofactors into pGEX-KG plasmids, creating GST-cofactor fusions *(43)*. Transform the constructs into the BL-21 strain of *E. coli* and identify the correct transformants by minipreparation and restriction cleavage analysis.

3.4.1.2. PREPARATION OF GST-COFACTOR FUSION PROTEINS *(43)*

1. Innoculate 0.2 mL of an overnight culture of the *E. coli* transformant of interest into 50 mL of LB media containing 100 µg/mL ampicillin. Grow the culture at 37°C with vigorous aeration until an $O.D._{600nm}$ of 1.0 is obtained. Add IPTG to 0.5 m*M* and continue the incubation for an additional 2 to 3 h.
2. Collect the bacteria by centrifugation at 1500*g* for 15 min. Resuspend the pellet in 0.5 mL of cold PBST containing complete protease inhibitor. Chill on ice.
3. Disrupt the cell suspension by sonication with a Branson sonicator, using a microtip, an output setting of 7, and a 15-s burst at 50% duty cycle. Cool the tube on ice for 3 min and repeat.
4. Clarify the sonicates by centrifugation in a microfuge at 4°C for 10 min and store the clear supernatants at –80°C in 20–100-µL aliquots.

3.4.1.3. COUPLING THE GST-COFACTORS TO GLUTATHIONE AGAROSE

1. Hydrate the glutathione–agarose (Sigma, St. Louis, MO) by swelling 0.2 g of the dry powder in 10 mL of PBST at 4°C for 20 min. Collect the glutathione agarose by a 15-s centrifugation at 1000*g*. Wash the glutathione agarose with 3 changes of 10 mL of 1X PBST and resuspend as a 50% slurry in the same buffer.
2. Thaw the frozen clarified *E. coli* sonicates. Pellet and discard any cryoprecipitate by a 5-min centrifugation (12,000*g*) in a microfuge at 4°C. Mix the resulting, clarified sonicates with glutathione agarose beads in 500 µL of PBST (containing Complete protease inhibitor). Incubate for 1 h at 4°C with constant gentle mixing (we use an inverting aliquot mixer). The precise amount of each lysate used per binding reaction must be determined empirically. As a starting point, we recommend using 5–10 µL of each sonicate per 15 µL of glutathione agarose (packed volume) for each binding reaction.
3. Collect the glutathione agarose, now containing the immobilized GST-cofactor, by a 10-s centrifugation in a microfuge and discard the supernatant. Wash the protein–agarose with 1 mL of cold PBST 3× and 1× with 1 mL of cold HEMG

buffer. The beads should be vortex mixed vigorously for 5 s between the washes. Resuspend as a 50% slurry in HEMG buffer.

3.4.1.4. BINDING OF THE RECEPTORS TO GST-COFACTOR–AGAROSE MATRIX

1. Prepare ^{35}S-labeled TRβ (wild-type or RTH mutant) by a coupled in vitro transcription and translation procedure (**Subheading 3.2.1.**). For each assay, mix 5–10 μL of the ^{35}S-labeled TRβ, 500 μL of HEMG (supplemented with 10 mg/mL BSA and complete protease inhibitor), and 15 μL (packed volume) of GST-cofactor immobilized on glutathione agarose. T3 hormone, or an equivalent amount of ethanol carrier, can also be added as desired.
2. Incubate the tubes for 1 h at 4°C with gentle mixing (we use an inverting aliquot mixer).
3. Collect the glutathione agarose, and any proteins bound to it, by a 10-s centrifugation in a microfuge. Wash the immobilized protein–agarose matrix 4× with 1 mL each of HEMG buffer, elute the proteins in 30 μL of 10 mM glutathione in 50 mM Tris-HCl, pH 8.0, and subject the eluate to SDS-PAGE and phosphorimager analysis. Compare the amount of radiolabeled TRβ bound to the GST-cofactor to the amount of radiolabeled TRβ used as input. Utilized nonrecombinant GST protein as a negative control in parallel experiments. Binding of the wild-type TRβ to nonrecombinant GST is less than 0.1% of input, whereas binding of greater than 1% of input TRβ to a GST-cofactor is usually diagnostic of a significant interaction.
4. A representative GST-pull-down experiment is shown in **Fig. 4**; note that both the wild-type and RTH mutant TRβ proteins used in this assay bind to GST-SMRT corepressor in the absence of hormone. The wild-type receptor releases from SMRT on addition of T3 hormone, whereas the RTH mutant remains bound to the corepressor even at high hormone concentrations. This impaired release from corepressor is one important molecular basis of the dominant-negative properties of many of the RTH mutant TRs.

3.4.2. Mammalian 2-Hybrid Assay

The 2-hybrid assay is based upon the fact that many eukaryotic transcriptional activators such as GAL4 consist of two modular domains: the DNA-binding domain (DBD), which binds to a specific promoter sequence, and the activation domain (AD) that recruits the transcriptional machinery *(44)*. In the mammalian 2-hybrid assay, otherwise physically separate GAL4DBD and GAL4AD domains are brought together on the promoter of a reporter gene by the interaction of a fused "bait" protein and a fused "prey" protein; as a result, the protein–protein interaction is manifested as activation of a suitable reporter gene containing binding sites for the GAL4 transcription factor *(45)* (*see* **Notes 10** and **11**).

3.4.2.1. SUBCLONING

Subclone the receptor interaction domains (RIDs) of the cofactor of interest (coactivator or corepressor) in-frame into the pSG5-GAL4DBD expression

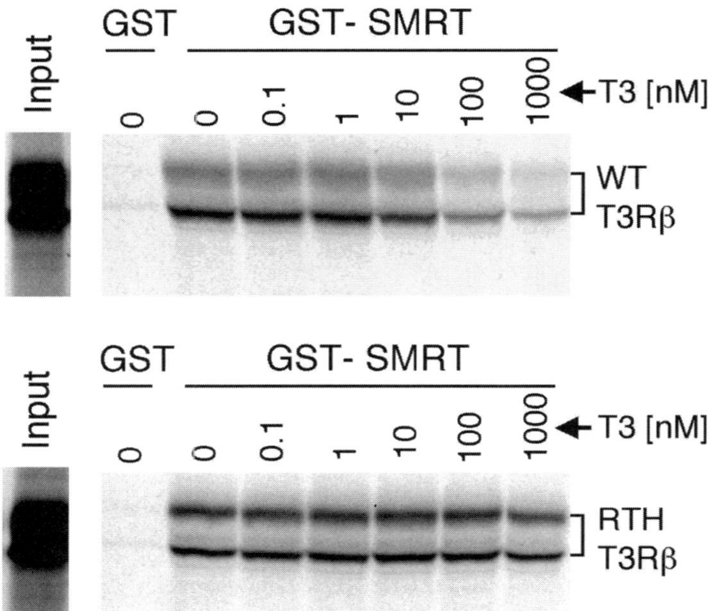

Fig. 4. Protein–protein interactions assayed by GST-pull-down method. Wild-type TRβ (top panel) and a RTH-TRβ mutant (bottom panel) were synthesized as ^{35}S-radiolabeled proteins by transcription–translation in vitro (input lanes). The radiolabeled receptors were then incubated with either GST-SMRT corepressor protein or with nonrecombinant GST protein, previously purified and immobilized on glutathione agarose (**Subheading 3.4.1.**). A range of hormone-concentrations were included in the binding reactions, as indicated. The immobilized protein preparations were subsequently washed, and any proteins bound to the GST or GST-SMRT preparations were eluted, resolved by SDS-PAGE, and then visualized by phosphorimager analysis. The positions of the wild-type and mutant receptor proteins are indicated to the right of the panel.

vector, creating GAL4DBD-cofactor fusions *(45)*. Similarly, subclone the TRβ allele of interest in frame into the pSG5-GAL4AD vector, creating a GAL4AD-receptor construct *(45)*. The C-terminal hormone binding domain of TRβ is sufficient to elicit an interaction with most corepressors and coactivators characterized.

3.4.2.2. Transient Transfection

1. Plate out CV-1 cells at a density of 5×10^4 cells/well in 12-well culture plates in DMEM supplemented with 0.1% bicarbonate and 10% FBS. Maintain the cells at 37°C in a 5% CO_2 atmosphere. Mix 25 ng of the appropriate pSG5-GAL4DBD-

cofactor plasmid, 100 ng of the appropriate pSG5-GAL4AD-receptor plasmid, 100 ng of the pGL-GAL4-17-mer luciferase reporter, 100 ng of the pCMV-lacZ internal control, and sufficient empty pSG5 vector to bring the total DNA to 500 ng/well. Use this mixture in the Lipofectin transfection protocol recommended by the manufacturer.

2. Twenty-four h after transfection, add either the desired amount of T3 hormone, or an equivalent vol of ethanol carrier, to the cells and incubate the cells for an additional 24 h at 37°C.

3. Harvest the cells, lyse in 1X lysis buffer and determine the luciferase and β-galactosidase activity. We use a luciferase assay kit (Promega) and determine β-galactosidase activity by colorimetric assay *(46)*. Normalize the luciferase activity to the β-galactosidase activity. The resulting relative luciferase activity is a measure of the extent of interaction between cofactor and receptor.

3.4.3. Biotin/DNA Pull-Down Assay

Create a double-stranded TRE probe as for EMSA (**Fig. 3A**), but modify it by incorporating biotin at the 5' ends of the oligonucleotides during the chemical synthesis. These complementary TRE oligonucleotides are annealed as described in **Subheading 3.3.2.** and are used as follows:

1. Mix 0.05–0.5 μg of the biotin-tagged annealed TRE probes with approx 250 ng of the TRβ receptor of interest (or a mixture of TRβ and RXRs when heterodimers are to be studied) in 500 μL of PBST containing 10 mg/mL BSA and complete protease inhibitor. We typically use receptors obtained from recombinant baculovirus/Sf9 cell preparations, as in **Subheading 3.3.1.** Add 15 μL (packed volume) of streptavidin-conjugated agarose beads (prewashed in PBST buffer) and incubate with constant gently mixing for 30 min at 4°C. Collect the protein/DNA complexes by a 10-s centrifugation in a microfuge, then vigorously wash the complexes 3× with 1 mL each of PBST buffer and 1× with 1 mL of HEMG buffer.

2. Incubate the resulting immobilized protein/DNA complexes from **step 1** for 1 h at 4°C with 5 μL of the appropriate ^{35}S-labeled cofactor protein (corepressor or coactivator) in 500 μL of HEMG buffer containing 10 mg/mL BSA and the desired amount of T3 hormone (or ethanol carrier). Continue the gentle mixing of the samples at 4°C.

3. Wash the immobilized DNA/protein complexes 4× with 1 mL each of HEMG buffer, and elute the proteins by boiling in SDS sample buffer. Resolve the proteins by SDS-PAGE. Visualize (and quantify) the radiolabeled cofactors that bound to the protein/DNA complex by autoradiography or phosphorimager analysis.

4. If desired, the amount and stoichiometry of receptor proteins in the DNA complex can also be determined by immunoblotting the eluted samples with antibodies against the receptors.

5. As a negative control, nuclear extracts of Sf9 cells infected with nonrecombinant baculovirus should be used in the place of the receptor extracts.

4. Notes

1. Identification of the mutations responsible for RTH syndrome is complicated by the fact that the vast majority of these patients are heterozygous for the genetic lesion. If direct sequence analysis of PCR products is performed, the RTH-TRβ lesion will appear as a double sequence at the point of heterozygosity (if a bp substitution), or as an extended region of double sequence (if a bp deletion or insertion). Similarly, if the PCR products are subcloned into plasmid vectors prior to sequence analysis, multiple clones will need to be analyzed to insure that both wild-type and potential mutant alleles are characterized.

2. The suggested primer annealing temperatures and $MgCl_2$ concentrations for the PCR reactions may need to be adapted for the particular primer and template sequences used.

3. There is a volatile component to most ^{35}S-amino acid preparations that can contaminate the work area if caution is not taken. Follow the recommendations of the manufacturer in regard to working with these isotopes.

4. The protease protection assay can also be performed in a reciprocal manner by holding the hormone concentration constant, but by varying the protease concentration. This alternative approach can be used as a probe of subtle differences in the conformations of wild-type vs mutant TRs, or can be used initially to determine an optimal protease concentration prior to use of the protease as a probe of hormone affinity. Siliconized tubes may be enhance the accuracy of this technique by preventing hormone adsorption on the hydrophobic walls of standard plastic tubes.

5. The advantage of using in vitro transcription and translation to produce the TRβ proteins for use in EMSA is that many different mutant receptors can be screened without the time-consuming process of subcloning and generating baculovirus stocks. Nevertheless, because the levels of protein expression by the in vitro translation system are low, receptors with weak DNA binding properties, such as is observed for TRβ homodimers, will be difficult to study with this method. Moreover, a nonspecific protein-TRE complex, apparently derived from polypeptides present in the reticulocyte lysate itself, often migrates at a mobility very similar to that of TRβ homodimers further complicating the analysis. In contrast to homodimers, heterodimers of TR and RXR have significantly higher affinities for DNA and typically can easily be used in EMSA in the form of in vitro translation products.

6. Three different RXR isoforms have been identified in vertebrates (RXRα, β, and γ). In most contexts examined to date, all three RXR isoforms display a similar ability to form heterodimers with TRβ and appear to function interchangeably in EMSA.

7. When TRβ protein synthesized by in vitro transcription and translation are used in the EMSA in place of the recombinant baculovirus/SF9 cell preparations, the following modifications are necessary:
   - Use 0.5X TBE and 1% glycerol in the 5% nondenaturing polyacrylamide gel.
   - Use up to 6 μL of in vitro translation products per assay in 20 μL of TnT

EMSA Buffer (60 mM HEPES, pH 7.8, 75 mM KCl, 7.5 mM MgCl$_2$, 1.5 μM ZnCl$_2$, 9% glycerol, 0.3 mg/mL BSA, 0.1 μg/μL poly-dI-dC).
- Use 4 μL of radiolabeled probe (100,000 cpm) per reaction.

8. We use thin elongated pipet tips or capillary tubes to underlay the samples under the electrode buffer when loading the EMSA gels. Use of a strong sidelight, and a contrasting visual background, permit monitoring of the loading due to the difference in density of the samples and the electrode buffer. Caution must be taken to avoid unnecessary mixing of the samples and the electrode buffer.

9. Significant amounts of the free probe can migrate out of the gel when the gels are dried. Some of this radioactive probe is absorbed and immobilized by the filter paper backing, but some radioactivity continues through the backing and can contaminate the gel dryer. To avoid this problem, we recommend using several additional layers of blotter paper and/or paper towels under the gel to absorb this potential contamination before it enters the body of the gel dryer itself.

10. It is important to use only the receptor interaction domains in the GAL4DBD fusions; inclusion of silencing domains or activation domains of the cofactors can lead to repression or activation of the reporter activity independent of the interaction with GAL4AD-receptors.

11. When analyzing the receptor–cofactor interaction by the mammalian 2-hybrid assay, the assay should be performed in a reciprocal fashion; the interactions between not only the GAL4-DBD-cofactor and GAL4-AD-receptor but also the GAL4-DBD-receptor and GAL4-AD-cofactor need to be tested.

References

1. Refetoff, S., Weiss, R. E., and Usala, S. J. (1993) The syndromes of resistance to thyroid hormone. *Endocr. Rev.* **14,** 348–399.
2. Kopp, P., Kitajima, K., and Jameson, J. L. (1996) Syndrome of resistance to thyroid hormone: insights into thyroid hormone action. *Proc. Soc. Exp. Biol. Med.* **211,** 49–61.
3. Chaterjee, V. K. K. (1997) Resistance to thyroid hormone. *Hormone Res.* **48,** 43 – 46.
4. Bodenner, D. L. and Lash, R. W. (1998) Thyroid disease mediated by molecular defects in cell surface and nuclear receptors. *Am. J. Med.* **105,** 524–538.
5. Mangelsdorf, D. J., Thummel, C., Beato, M., et al. (1995) The nuclear receptor superfamily: the second decade. *Cell* **83,** 835–839.
6. Apriletti, J. W., Ribeiro, R. C., Wagner, R. L., et al. (1998) Molecular and structural biology of thyroid hormone receptors. *Clin. Exp. Pharmacol. Physiol. Suppl.* **25,** S2–S11.
7. Zhang, J. and Lazar, M. A. (2000) The mechanism of action of thyroid hormones. *Annu. Rev. Physiol.* **62,** 439–466.
8. Damm, K., Thompson, C. C., and Evans, R. M. (1989) Protein encoded by v-erbA functions as a thyroid-hormone receptor antagonist. *Nature* **339,** 593–597.
9. Sap, J., Munoz, A., Schmitt, J., Stunnenberg, H., and Vennstrom, B. (1989) Repression of transcription mediated at a thyroid hormone response element by the v-erb-a oncogene product. *Nature* **340,** 242–244.

10. Baniahmad, A., Kohne, A. C., and Renkawitz, R. (1992) A transferable silencing domain is present in the thyroid hormone receptor, in the v-erb a oncogene product and in the retinoic acid receptor. *EMBO J.* **11,** 1015–1023.

11. Horwitz, K. B., Jackson, T. A., Bain, D. L., Richer, J. K., Takimoto, G. S., and Tung, L. (1996) Nuclear receptor coactivators and corepressors. *Mol. Endocrinol.* **10,** 1167–1177.

12. Shibata, H., Spencer, T. E., Oñate, S. A., et al. (1997) Role of co-activators and co-repressors in the mechanism of steroid/thyroid receptor action. *Recent Prog. Horm. Res.* **52,** 141–164.

13. Chen, J. D. and Li, H. (1998) Coactivation and corepression in transcriptional regulation by steroid/nuclear hormone receptors. *Crit. Rev. Eukaryot. Gene Expr.* **8,** 169–190.

14. Collingwood, T. N., Urnov, F. D., and Wolffe, A. P. (1999) Nuclear receptors: coactivators, corepressors and chromatin remodeling in the control of transcription. *J. Mol. Endocrinol.* **23,** 255–275.

15. Xu, L., Glass, C. K., and Rosenfeld, M. G. (1999) Coactivator and corepressor complexes in nuclear receptor function. *Curr. Opin. Genet. Dev.* **9,** 140–147.

16. Burke, L. J. and Baniahmad, A. (2000) Co-repressors 2000. *FASEB J.* **14,** 1876–1888.

17. Sakurai, A., Miyamoto, T., Refetoff, S., and DeGroot, L. J. (1990) Dominant negative transcriptional regulation by a mutant thyroid hormone receptor-beta in a family with generalized resistance to thyroid hormone. *Mol. Endocrinol.* **4,** 1988–1994.

18. Chatterjee, V. K., Nagaya, T., Madison, L. D., Datta, S., Rentoumis, A., and Jameson, J. L. (1991) Thyroid hormone resistance syndrome. Inhibition of normal receptor function by mutant thyroid hormone receptors. *J. Clin. Invest.* **87,** 1977–1984.

19. Baniahmad, A., Tsai, S. Y., O'Malley, B. W., and Tsai, M. J. (1992) Kindred S thyroid hormone receptor is an active and constitutive silencer and a repressor for thyroid hormone and retinoic acid responses. *Proc. Natl. Acad. Sci. USA* **89,** 10,633–10,637.

20. Collingwood, T. N., Adams, M., Tone, Y., and Chatterjee, V. K. (1994) Spectrum of transcriptional, dimerization, and dominant negative properties of twenty different mutant thyroid hormone beta-receptors in thyroid hormone resistance syndrome. *Mol. Endocrinol.* **8,** 1262–1277.

21. Nagaya, T., Fujieda, M., and Seo, H. (1998) Requirement of corepressor binding of thyroid hormone receptor mutants for dominant negative inhibition. *Biochem. Biophys. Res. Commun.* **247,** 620–623.

22. Yen, P. M. and Chin, W. W. (1994) Molecular mechanisms of dominant negative activity by nuclear hormone receptors. *Mol. Endocrinol.* **8,** 1450–1454.

23. Yoh, S. M., Chatterjee, V. K., and Privalsky, M. L. (1997) Thyroid hormone resistance syndrome manifests as an aberrant interaction between mutant T3 receptors and transcriptional corepressors. *Mol. Endocrinol.* **11,** 470–480.

24. Collingwood, T. N., Wagner, R., Matthews, C. H., et al. (1998) A role for helix 3 of the TRbeta ligand-binding domain in coactivator recruitment identified by characterization of a third cluster of mutations in resistance to thyroid hormone. *EMBO J.* **17,** 4760–4770.

25. Clifton-Bligh, R. J., de Zegher, F., Wagner, R. L., et al. (1998) A novel TR beta mutation (R383H) in resistance to thyroid hormone syndrome predominantly

impairs corepressor release and negative transcriptional regulation. *Mol. Endocrinol.* **12,** 609–621.

26. Safer, J. D., Cohen, R. N., Hollenberg, A. N., and Wondisford, F. E. (1998) Defective release of corepressor by hinge mutants of the thyroid hormone receptor found in patients with resistance to thyroid hormone. *J. Biol. Chem.* **273,** 30,175–30,182.
27. Tagami, T. and Jameson, J. L. (1998) Nuclear corepressors enhance the dominant negative activity of mutant receptors that cause resistance to thyroid hormone. *Endocrinology* **139,** 640–650.
28. Tagami, T., Gu, W. X., Peairs, P. T., West, B. L., and Jameson, J. L. (1998) A novel natural mutation in the thyroid hormone receptor defines a dual functional domain that exchanges nuclear receptor corepressors and coactivators. *Mol. Endocrinol.* **12,** 1888–1902.
29. Yoh, S. M. and Privalsky, M. L. (2000) Resistance to thyroid hormone (RTH) syndrome reveals novel determinants regulating interaction of T3 receptor with corepressor. *Mol. Cell Endocrinol.* **159,** 109–124.
30. Beck-Peccoz, P. and Chatterjee, V. K. K. (1994) The Variable Clinical Phenotype in Thyroid Hormone Resistance Syndrome. *Thyroid* **4,** 225–232.
31. Pohlenz, J., Wirth, S., Winterpacht, A., Wemme, H., Zabel, B., and Schonberger, W. (1995) Phenotypic variability in patients with generalised resistance to thyroid hormone. *J. Med. Genet.* **32,** 393–395.
32. Weiss, R. E., Hayashi, Y., Nagaya, T., Petty, K. J., Murata, Y., Tunca, H., Seo, H., and Refetoff, S. (1996) Dominant inheritance of resistance to thyroid hormone not linked to defects in the thyroid hormone receptor alpha or beta genes may be due to a defective cofactor. *J. Clin. Endocrinol. Metab.* **81,** 4196–4203.
33. Safer, J. D., Langlois, M. F., Cohen, R., et al. (1997) Isoform variable action among thyroid hormone receptor mutants provides insight into pituitary resistance to thyroid hormone. *Mol. Endocrinol.* **11,** 16–26.
34. Adams, M., Matthews, C., Collingwood, T. N., Tone, Y., Beck-Peccoz, P., and Chatterjee, K. K. (1994) Genetic analysis of 29 kindreds with generalized and pituitary resistance to thyroid hormone. Identification of thirteen novel mutations in the thyroid hormone receptor beta gene. *J. Clin. Invest.* **94,** 506–515.
35. Syvänen, A. C., Aalto-Setälä, K., Kontula, K., and Söderlund, H. (1989) Direct sequencing of affinity-captured amplified human DNA application to the detection of apolipoprotein E polymorphism. *FEBS Lett.* **258,** 71–74.
36. Chen, E. Y., and Seeburg, P. H. (1985) Supercoil sequencing: a fast and simple method for sequencing plasmid DNA. *DNA* **4,** 165–170.
37. Horton, R. M. (1993) In vitro recombination and mutagenesis of DNA, in *Methods in Molecular Biology*, Vol. 15 (White, B.A. ed.), Humana Press, Totowa, NJ, pp. 251– 261.
38. Inoue, A., Yamakawa, J., Yukioka, M., and Morisawa, S. (1983) Filter-binding assay procedure for thyroid hormone receptors. *Anal. Biochem.* **134,** 176–183.
39. Weinberger, C., Thompson, C. C., Ong, E. S., Lebo, R., Gruol, D. J., and Evans, R. M. (1986) The c-erb-A gene encodes a thyroid hormone receptor. *Nature* **324,** 641–646.

40. Lin, B. C., Hong, S. H., Krig, S., Yoh, S. M., and Privalsky, M. L. (1997) A conformational switch in nuclear hormone receptors is involved in coupling hormone binding to corepressor release. *Mol. Cell Biol.* **17,** 6131–6138.
41. Lane, D., Prentki, P., and Chandler, M. (1992) Use of gel retardation to analyze protein-nucleic acid interactions. *Microbiol. Rev.* **56,** 509–528.
42. Forman, B. M., Casanova, J., Raaka, B. M., Ghysdael, J., and Samuels, H. H. (1992) Half-site spacing and orientation determines whether thyroid hormone and retinoic acid receptors and related factors bind to DNA response elements as monomers, homodimers, or heterodimers. *Mol. Endocrinol.* **6,** 429–442.
43. Guan, K. L. and Dixon, J. E. (1991) Eukaryotic proteins expressed in *Escherichia coli*: an improved thrombin cleavage and purification procedure of fusion proteins with glutathione S-transferase. *Anal. Biochem.* **192,** 262–267.
44. Ma, J. and Ptashne, M. (1987) A new class of yeast transcriptional activators. *Cell* **51,** 113–119.
45. Fearon, E. R., Finkel, T., Gillison, M. L., et al. (1992) Karyoplasmic interaction selection strategy: a general strategy to detect protein-protein interactions in mammalian cells. *Proc. Natl. Acad. Sci. USA* **89,** 7958–7962.
46. Vivanco, M. D., Johnson, R., Galante, P. E., Hanahan, D., and Yamamoto, K. R. (1995) A transition in transcriptional activation by the glucocorticoid and retinoic acid receptors at the tumor stage of dermal fibrosarcoma development. *EMBO J.* **14,** 2217–2228.

9

Function of Thyroid Hormone Receptors During Amphibian Development

Sashko Damjanovski, Laurent M. Sachs, and Yun-Bo Shi

1. Introduction

Thyroid hormone (T3) plays important roles during vertebrate development *(1)*. In humans, T3 is detected in the embryonic plasma by 6 mo and rises to high levels around birth *(2)*. During this postembryonic period, extensive tissue remodeling and organogenesis take place. T3 deficiency during human development leads to developmental defects, such as mental retardation, short stature, and in the most severe form, cretinism *(1,3)*. Likewise, T3 is also critical for amphibian development. It is the controlling agent of anuran metamorphosis, a process that transforms a tadpole into a tailless frog *(1,4)*. Blocking synthesis of endogenous T3 leads to the formation of giant tadpoles that cannot metamorphose, while addition of exogenous T3 to premetamorphic tadpoles causes precocious metamorphosis. Importantly, most, if not all, organs are genetically predetermined to undergo specific changes, and these changes are organ autonomous. Thus, T3 appears to act directly on individual metamorphosing organs. Such properties have made anuran metamorphosis one of the best-studied postembryonic developmental process at morphological, cellular, and biochemical levels and paved the way for current molecular investigations of the underlying mechanisms.

1.1. Gene Regulation by Thyroid Hormone Receptors

Thyroid hormone functions by regulating gene expression through thyroid hormone receptors (TRs). TRs are DNA-binding transcription factors that belong to the steroid hormone receptor superfamily *(5–7)*. Like most other members of this family, TRs consist of several distinct domains, including the

From: *Methods in Molecular Biology, Vol. 202: Thyroid Hormone Receptors: Methods and Protocols*
Edited by: A. Baniahmad © Humana Press Inc., Totowa, NJ

DNA-binding domain in the N terminal half of the protein and a hormone-binding domain in the C terminal half of the protein.

TRs regulate gene expression mainly as heterodimers with 9-cis retinoid x receptors (RXRs) *(5–9)*. TR/RXR heterodimers are nuclear proteins that bind constitutively to thyroid hormone response elements (TREs) in chromatin *(10,11)*. Current understanding suggests that TR/RXR has dual functions: repressing target gene expression in the absence of T3 and activating it when T3 is present *(9,12)*.

Both transcriptional repression and activation appear to be mediated by multicomponent cofactor complexes. In the absence of thyroid hormone, TR/RXR binds to corepressor such as NCoR, SMRT, SUN-CoR, and Alien *(13–20)*. Multiple corepressor complexes have been shown to contain histone deacetylases (HDACs) such as Rpd3. Thus transcriptional repression by unliganded TR/RXR may be mediated, in part, through the recruitment of HDAC complexes leading to remodeling chromatin. Consistent with these finding, we have shown that in the frog oocyte, TR/RXR-mediated repression can be eliminated by blocking HDAC activity with the drug trichostatin A (TSA), while repression by overexpressed Rpd3 can be reversed by thyroid hormone-bound TR/RXR on a chromatinized thyroid hormone-responsive promoter *(19,21,22)*.

The addition of T3 leads to the release of corepressors and concurrent recruitment of coactivators. Many TR-binding coactivators such as SRC-1 and CBP/p300 are themselves histone acetyltransferases or acetylases (HATs) and also form multicomponent complexes *(14,18,20,23)*. Thus, thyroid hormone-binding to TR may activate transcription, at least in part, through increasing histone acetylation.

Transcriptional regulation by TR/RXR is, however, much more complex than the simple involvement of histone acetylation. For instance, TR/RXR can also interact directly with basal transcriptional machinery *(7)*. Moreover, at least one additional coactivator complex, the DRIP/TRAP complex, does not contain HAT activity. It contains many subunits of the holo-RNA polymerase complex, suggesting that TR/RXR may also regulate transcription by directly influencing the transcriptional machinery *(23)*. Finally, using a reconstituted in vivo transcriptional system in the frog oocyte, we have shown that transcriptional activation by TR/RXR in the presence of T3 leads to chromatin disruption *(11)*. The mechanism for this drastic chromatin remodeling is not clear at the present time, but does not appear to be caused by changes in histone acetylation levels *(22)*. Thus, the mechanism of transcriptional regulation by TR/RXR may vary depending upon the target gene and the target tissue, where the levels of different cofactors may vary. It can be a 2-step process. The first step is chromatin remodeling, including the disruption of chromatin structure and/or histone acetylation through the recruitment of HAT/HDAC complexes and/or chromatin remodeling complexes. Then transcriptional activation takes place through direct interactions with the RNA polymerase complex, e.g., by means of the DRIP/TRAP complex. An alternative, but not necessarily mutually exclusive, model would be that TR/RXR may regulate transcription through parallel pathways, including both chromatin remodeling and direct interactions with the transcription machinery *(23,24)*.

1.2. Expression of TR and RXR in Xenopus laevis and Its Implications in Frog Development

Four *TR* genes, two *TRα*, and two *TRβ* genes, are present in *Xenopus laevis (25)*. The total dependence of anuran metamorphosis on T3 offers an opportunity to study TR/RXR function during development. As expected, both *TRα* and *TRβ* genes are highly expressed during metamorphosis (**Fig. 1**) *(25,26)*. In addition, *RXR* genes are also expressed during metamorphosis. More importantly, the expression of both *TR* and *RXR* genes correlates temporally with metamorphosis of individual organs *(26)*. Thus, high levels of both TR and RXR mRNAs are present in the limbs during the early stages of metamorphosis (stage 54–58), when limb morphogenesis takes place. Subsequently, as the limbs undergo growth with little morphological change, both *TR* and *RXR* genes are down-regulated. On the other hand, both *TR* and *RXR* genes are up-regulated in the tail toward the end of metamorphosis (after stage 60), which corresponds to the period of tail resorption. Such correlation argue for TR/RXR heterodimers to be the mediators of the controlling effects of T3 on metamorphosis in all organs *(27)*.

Interestingly, *TRα* and *TRβ* genes are differentially regulated during development. The *TRβ* genes have little expression prior to metamorphosis, but are themselves direct thyroid hormone-response genes (**Fig. 1**) *(25,28,29)*. Their expression is up-regulated by the rising concentration of endogenous T3 during metamorphosis. In contrast, the TRα genes are activated shortly after the end of embryogenesis, and their mRNAs reach high levels by Stage 45, when tadpole feeding begins.

The expression profiles together with the ability of TR to both repress and activate thyroid hormone-inducible genes in the absence and presence of thyroid hormone, respectively, suggest dual functions for TRs during development *(24)*. That is, in premetamorphic tadpoles, TRs, mostly TRα, act to repress thyroid hormone response genes. As many of these genes are likely to participate in metamorphosis *(30)*, their repression by unliganded TR will help to prevent premature metamorphosis and ensure a proper period of tadpole growth. As the thyroid gland matures, T3 is synthesized and secreted into the plasma to transform TRs from repressors to activators, which will induce the

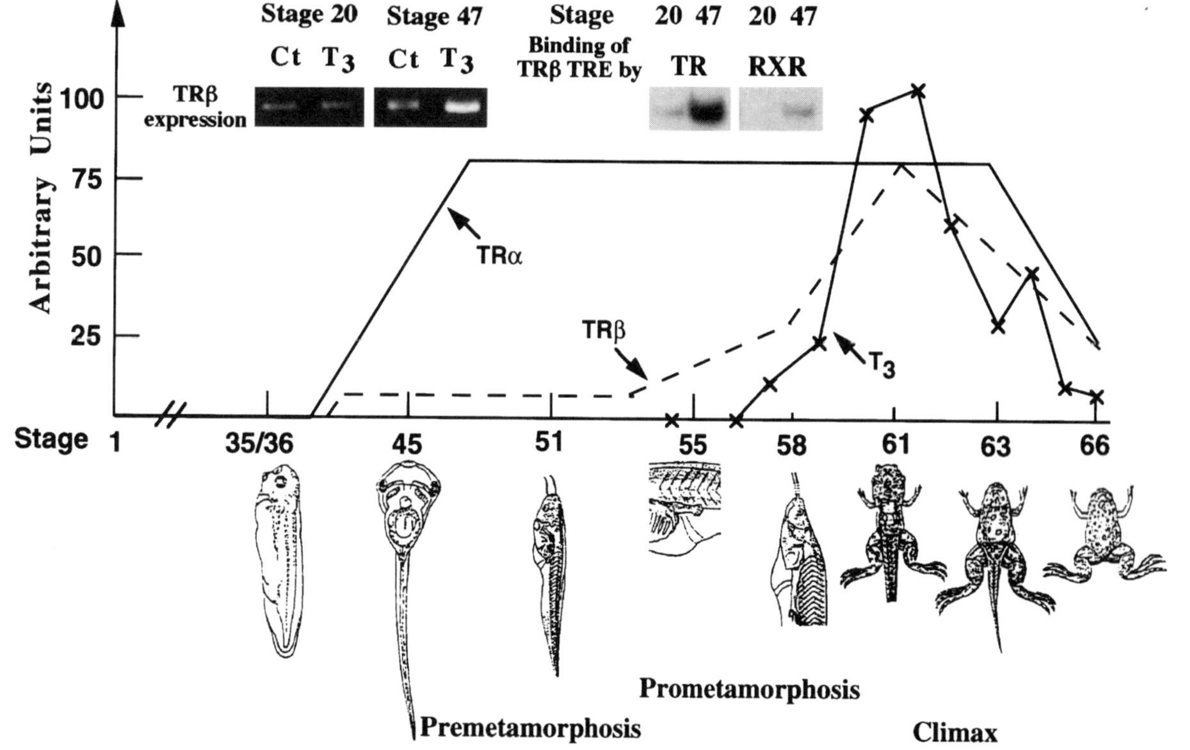

Fig. 1. Stage-dependent TR expression and binding to TREs during frog development. The line graphs show the TR mRNA levels and plasma T3 concentrations during *Xenopus laevis* development, based on Yaoita and Brown (1990) and Leloup and Buscaglia (1977), respectively *(33,44)*. The insert on the top left shows that TRβ is expressed at low levels in premetamorphic tadpoles (lanes labeled with Ct) and can be induced by treating premetamorphic tadpoles with T3 (lanes labeled with T3) at stage 47 when TRα is expressed, but not embryos at stage 20 when there is little TR *(32)*. On the top right, ChIP assays indicate that the binding of TR and RXR to the TRE in the *TRβ* gene in the *Xenopus laevis* tadpoles, but not embryos, is correlated with the high levels of TRα expression at stage 47 *(32)*.

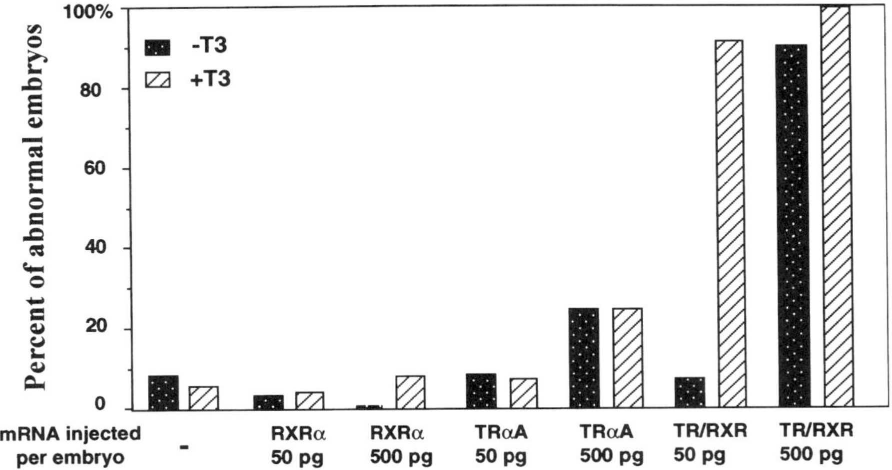

Fig. 2. RXR is critical for TR to mediate the developmental effects of T3. Embryos injected with indicated amounts of mRNAs, for TR or RXR or both, were cultured in the presence or absence of 100 n*M* T3 and analyzed for phenotypes 48 h after fertilization. The percentages of the embryos that were deformed are plotted. The phenotypes were different dependent upon the presence or absence of T3 (*see* Puzianowsak-Kuznicka et al. [1997] for more details *[8]*).

expression of T3 response genes, including the *TRβ* genes, thus leading to metamorphosis.

1.3. Mechanistic and Functional Analyses of TR Action During Development

To test the dual function hypothesis above, we have made use of the lack of or the very low levels of TRs and RXRs during embryogenesis and the ability to introduce exogenous proteins into embryos by microinjecting their mRNAs into fertilized eggs *(8)*. Overexpression of TRα or TRβ or RXRα alone has little effect on embryonic development both in the presence or absence of T3 (**Fig. 2**). On the other hand, co-expression of any TR with RXR causes developmental abnormalities in embryos, but the phenotypes of the embryos are distinct depending upon the levels of the overexpression of TR/RXR, and whether T3 is present or not (**Fig. 2**). This is in agreement with the fact that TR/RXR heterodimers are repressors in the absence of T3, but activators in the presence of the hormone. More importantly, by analyzing the expression of known T3 response genes, we have shown that TR/RXR, but neither receptor alone, regulates T3 response genes efficiently and specifically in the embryos, repressing them in the absence of T3, but activating them when T3 is present *(8)*.

If TR/RXR indeed functions to repress T3 response genes in premetamorphic tadpoles, we would expect that they are bound to TREs of endogenous T3 response genes independent of T3. To prove this, we have carried out chromatin immunoprecipitation (ChIP) assays using antibodies against TR or RXR *(24)*. Polymerase chain reaction (PCR) analysis of the TRE regions of two known T3 response genes, the *TRβ* and *TH/bZip* genes (the only two frog genes whose TREs have been characterized) *(29,31)* on precipitated embryonic or tadpole chromatin fragments, has revealed that little TRs or RXRs are present at the TREs in embryos when there is little TR expressed, but both TREs are bound by TR/RXR in tadpoles when TRs (at least TRα) are expressed (**Fig. 1**) *(32)*. This agrees well with the ability of T3 to induce the expression of T3 response genes in tadpoles but not embryos (**Fig. 1**) *(33)*. Furthermore, both T3 response genes are known to be inducible ubiquitously in different tissues (**Fig. 3A**) *(33,34)*. Consistently, ChIP assays on individual organs have shown that the TREs of both thyroid hormone-response genes are bound by TR/RXR in different organs (**Fig. 3B**) *(32)*.

The presence of TR, but not T3, in premetamorphic tadpoles offers an opportunity to investigate the involvement of HDACs in gene repression by TR in vivo. Thus, we have treated embryos and tadpoles with TSA which is a specific chemical inhibitor for histone deacetylases *(32)*. Analysis of the expression of T3 response genes have shown that TSA induces precocious expression of T3 response genes in the intestine and tail (**Fig. 3A**) *(32)*, supporting the involvement of histone deacetylase in the repression of T3 response genes by unliganded TR/RXR.

ChIP assays using an antibody against acetylated histone H4 have shown that T3 treatment of premetamorphic tadpoles leads to an increase in histone acetylation specifically at the TRE regions of T3 response genes in intestine and tail without affecting global histone acetylation (**Fig. 4**) *(32)*. Similarly, TSA treatment of premetamorphic tadpoles elevates histone acetylation levels of the TRE regions of T3 response genes, as well as global histone acetylation levels in the intestine and tail. These results together argue strongly for a role of histone deacetylase in gene repression by unliganded TR/RXR, at least in the intestine and tail of premetamorphic tadpoles.

Interestingly, ChIP analyses have shown that T3 does not induce any increase in histone acetylation of the TRE regions in whole tadpoles, even though it reduces the association of histone deacetylase Rpd3 with the TRE regions *(32)*. In addition, little up-regulation of the T3 response genes by TSA treatment can be detected when gene expression is analyzed on whole animals. Thus, there are likely tissue-specific mechanisms involving varying degrees of participation of histone acetylation in gene regulation by TR/RXR *(32)*.

In summary, *TR* and *RXR* genes are coordinately expressed during frog development, and their expression is correlated with organ-specific changes

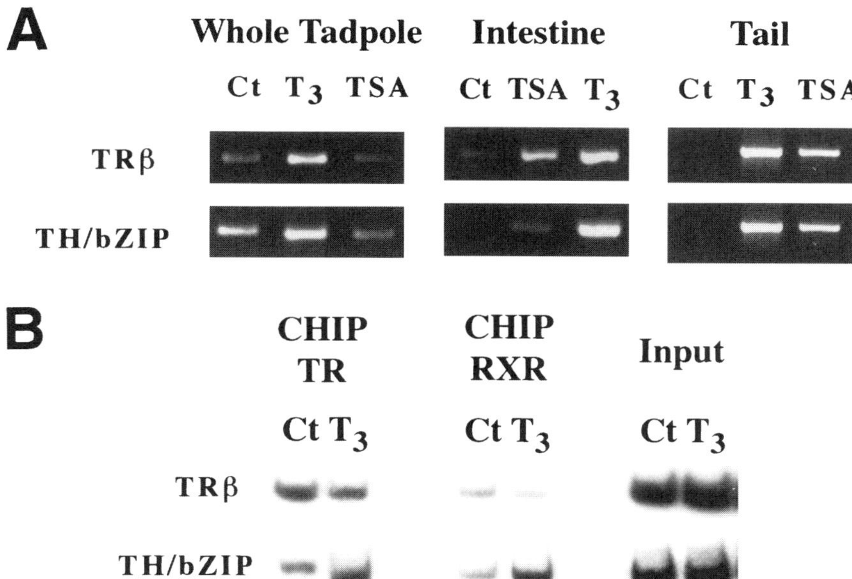

Fig. 3. Induction of the transcription of T3 response genes, *TRβ* and *TH/bZIP*, by T3 and TSA and constitutive binding of TR/RXR to TRE in premetamorphic tadpoles *(32)*. (**A**) Differential effects of T3 and TSA on gene expression. 55 tadpoles were treated for 2 d with T3 (10 n*M*) or TSA (100 n*M*). Total RNA was extracted from whole animals, intestine or tail and used for PCR analysis of *TRβ* and *TH/bZIP* expression. Note that T3 treatment increased mRNA levels of T3 response genes in whole animals, intestine, and tail, while TSA treatment altered T3, response gene expression only in intestine and tail. (**B**) TR/RXR binds to TREs in chromatin constitutively. Chromatin isolated from tail of stage 55 tadpoles treated with or without T3, was immunoprecipitated with antibodies against TR or RXR and analyzed by PCR. Note that TR and RXR are bound to the TREs of both genes with or without T3 treatment. The input control was obtained by PCR on DNA prior to immunoprecipitation. Identical results were obtained from analysis on the intestine or whole tadpoles.

during metamorphosis *(27)*. In premetamorphic tadpoles, TRs (mainly TRα) function as unliganded TR/RXR heterodimers. They bind to the TREs in target genes to repress their expression, most likely involving histone deacetylases. As at least some of these T3 response genes participate in metamorphosis, their repression by TR may be required to prevent premature metamorphosis and ensure a proper period of tadpole growth. As endogenous T3 becomes available during metamorphosis, TRs bind to T3 and become transcriptional activators. This leads to chromatin remodeling, including changes in local histone acetylation levels, and induces the expression of T3 response genes and, thus,

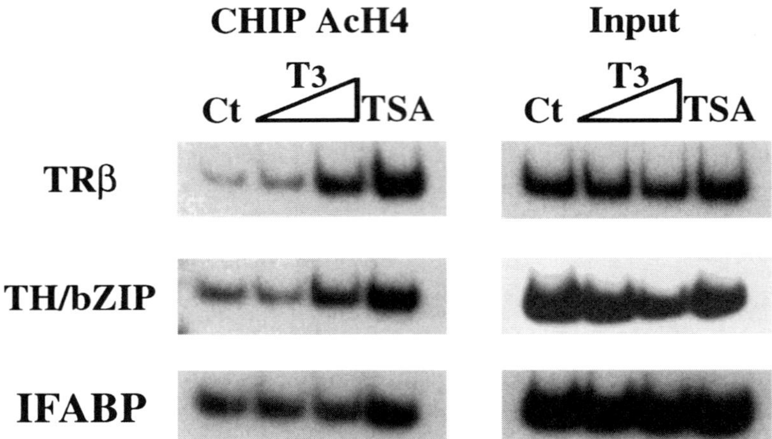

Fig. 4. T3 treatment increases histone H4 acetylation specifically at the TRE regions of T3 response genes in premetamorphic tadpole intestine *(32)*. Stage 55 tadpoles were treated for 2 d with T3 (10 n*M*) or TSA (100 n*M*). Nuclei extract from the intestine were used for the ChIP assay using an antibody against acetylated histone H4. Aliquots of the chromatin prior to immunoprecipitation were used in PCR as a DNA control (Input). Note that T3 treatment led to the increase in histone acetylation of the TRE regions of the T3 response genes, TRß and TH/bZIP, but not the intestine-specific IFABP gene promoter, which is not directly regulated by T3. In contrast, TSA elevated acetylation levels of all 3 genes.

activating the gene expression cascades that controls metamorphosis. The future challenge will be to determine the nature of chromatin remodeling and the roles of various TR-interacting factors/complexes in gene regulation by TRs during development. Finally, transgenesis using sperm mediated gene transfer in *Xenopus (35)* offers a direct means to study receptor function and its underlying mechanism during metamorphosis *(36)*.

2. Materials
2.1. Analysis of TR Function During Development Through mRNA Injection into Fertilized Eggs

1. Adult *Xenopus laevis*, Nasco (Wisconsin).
2. Needle puller, World Precision Instruments, PUL-1 (*see* **Note 1**).
3. Needles, Drummond Microcaps 1-000-0300 (*see* **Note 1**).
4. Injector, Eppendorf FemtoJet Microinjector (*see* **Note 1**).
5. Thyroid Hormone T3, Sigma T-2752.
6. TR and RXR expression constructs are as described in Purianowska-Kuznicka et al. (1997) *(8)*.
7. RNA transcription kit, Ambion mMessage mMachine.

2.2. Use of Chromatin Immunoprecipitation for Investigating TR Action in Development

1. Adults and premetamorphic tadpoles of the South African clawed frog *Xenopus laevis* were obtained from Nasco (Wisconsin). Embryo were prepared by in vitro fertilization as described *(8)*. Developmental stages were determined according to Nieuwkoop and Faber (1956) *(37)*.
2. Nuclei extraction buffer A: 2.2 M sucrose, 10 mM Tris-HCl, pH 7.5, 3 mM CaCl$_2$, 0.5% Triton X-100, 1 mM phenylmethylsulfonyl fluoride (PMSF), 5 µg/mL aprotinin, 1 µg/mL pepstatin A. Buffer A should be made just before use.
3. Nuclei extraction buffer B: 0.25 M sucrose, 10 mM Tris-HCl, pH 7.5, 3 mM CaCl$_2$, 1 mM PMSF, 5 µg/mL aprotinin, 1 µg/mL pepstatin A. Buffer B should be made just before use.
4. Lysis buffer: 1% sodium dodecyl sulfate (SDS), 10 mM EDTA, 50 mM Tris-HCl, pH 8.1, 1 mM PMSF, 1 µg/mL aprotinin, 1 µg/mL pepstatin A. Add protease inhibitors just before use.
5. ChIP dilution buffer: 0.01% SDS, 1.1% Triton X-100, 1.2 mM EDTA, 16.7 mM Tris-HCl, pH 8.1, 167 mM NaCl, 1 mM PMSF, 1 µg/mL aprotinin, 1 µg/mL pepstatin A. Add protease inhibitors just before use.
6. Salmon sperm DNA/protein A agarose slurry: 500 µL packed beads, 200 µg sonicated salmon sperm DNA, 500 µg bovine serum albumin (BSA), 1.5 mg recombinant protein A as a 50% gel slurry in Tris-EDTA (TE) buffer containing 0.05% sodium azide.
7. TE: 10 mM Tris-HCl, 1 mM EDTA, pH 8.0.
8. Antibodies: anti-TR and anti-RXR antibody *(26)*, anti-Rpd3 antibody *(32)*, and anti-acetylated-H4 (Upstate Biotechnology, Lake Placid, USA).
9. Low salt complex wash buffer: 0.1% SDS, 1% Triton X-100, 2 mM EDTA, 20 mM Tris-HCl, pH 8.1, 150 mM NaCl.
10. High salt complex wash buffer: 0.1% SDS, 1% Triton X-100, 2 mM EDTA, 20 mM Tris-HCl, pH 8.1, 500 mM NaCl.
11. LiCl immune complex wash buffer: 0.25 M LiCl, 1% Nonidet P-40 (NP40), 1% deoxycholate, 1 mM EDTA, 10 mM Tris-HCl, pH 8.1.
12. Elution buffer: 1% SDS, 0.1 M NaHCO$_3$. Prepare just before use.
13. PCR materials buff, oligonucleitide, dNTP (Takara ExTaq, Intergene): The primers used 3' *(32)* for TRβ promoter: forward 5'-GTAAGCTGCCTGTGTCTATAC-3' and reverse 5'-GACAGTCAGAGGAACTG-3'; for TH/bZIP promoter: forward 5'-TCTCCCTGTTGTGTATAATGG-3' and reverse 5'-CT CCCAA CCCTACAGAGTTCA-3'; for a segment of TRβ transcribed sequence: forward 5'-CAGAAACCTGAACCCACACAA-3' and reverse 5' CACTTTTCCACCCT CGGGCGCATT-3' (located respectively in exons 3 and 4) *(32)*; and for IFABP promoter: forward 5'-ATAGCAGCAGGTGGTTGCG-3' and reverse 5'-GGCCA CAAGATCTACTCG-3' *(32)*.
14. Other reagents: 3,5,3' triiodothyronine (T3; Sigma, cat. no. T-2752), trichostatin A (Wako, cat no. 204-11991), Liebovitz medium L15 (Life technology), 37% formaldehyde (*see* **Note 2**)

2.3. Transgenic Analysis of TR Function During Development

1. 1X Nuclear preparation buffer (NPB): 250 mM sucrose (1.5 M stock, filtered and aliquoted at $-20°$C), 15 mM HEPES (1 M stock, adjusted with KOH, so that the pH of the 15 mM buffer is 7.7, stored at $-20°$C), 1 mM EDTA, 0.5 mM spermidine trihydrochloride (Sigma, cat no. S-2501; 10 mM stock, filtered aliquots stored at $-20°$C), 0.2 mM spermidine tetrahydrochloride (Sigma, cat. no. D-1141;, 10 mM stock, filtered aliquots stored at $-20°$C), 1 mM dithiothreitol (DTT) (Sigma, cat no. D-0632; 100 mM stock, filtered aliquots stored at $-20°$C).
2. 1X Marc's modified ringers (MMR): 100 mM NaCl, 2 mM KCl, 1 mM MgCl$_2$, 2 mM CaCl$_2$, 5 mM HEPES, pH 7.5.
3. Digitonin: 5 µL of 10mg/mL digitonin in dimethyl sulfoxide (DMSO) (Sigma, cat. no. D-5628), stored frozen.
4. BSA: 10% BSA (fraction V; Sigma, cat. no. A-7906, in water, adjusted to pH 7.6 with KOH, aliquoted to 1 mL and stored at $-20°$C.
5. Sperm dilution buffer (sdb): 250 mM sucrose, 75 mM KCl, 0.5 mM spermidine trihydrochloride.
6. 0.2 mM Spermidine tetrahydrochloride, add about 80 µL of 0.1 N NaOH per 20 mLs to pH 7.3–7.5, 0.5 mL aliquots stored at $-20°$C.
7. Hoechst No. 33342: (Sigma, cat. no. B-2261) 10 mg/mL in water, stored light tight at $-20°$C.
8. 20X Extract buffer (XB) stock: 2 M KCl, 20 mM MgCl$_2$, 2 mM EGTA, filtered and stored at $-20°$C
9. Extract buffer (make about 100mL): 1X XB (from 20X stock), 50 mM sucrose from 1.5 M stock, filter, store at $-20°$C, 10 mM HEPES (1 M stock pH adjusted with 5.5 mL 10 N KOH/100 mL, filtered, and stored at $-20°$C) final pH (10 mM) is 7.7.
10. 2.5% Cysteine: made fresh in 1X MMR, adjusted to pH 7.8 with NaOH.
11. Crude cytostatic factor (CSF)-XB: 10 mM KCl, 0.1 mM CaCl$_2$, 1 mM MgCl$_2$, 10 mM HEPES, pH 7.7, 50 mM sucrose, 5 mM EGTA, pH 7.7.
12. Energy mixture: 150 mM creatine phosphate (Roche, cat no. 127 574), 20 mM ATP (Roche, 519 979), 20 mM MgCl$_2$, stored in 0.1 mL aliquots at $-20°$C.
13. Pregnant-mare serum gonadotropin (PMSG): 100 U/mL in water, stored at $-20°$C (Calbiochem, cat. no. 367222).
14. Chorionic Gonadotropin (HCG): 1000 U/mL in water, stored at 4°C (Sigma, cat. no. CG-10).
15. 0.6X MMR plus 6% Ficoll: used prior to gastrulation.
16. 0.1X MMR plus 50 µg/mL gentamicin: used after gastrulation.
17. Linearized plasmids for transgenesis: 200–250 ng/ µL. *Xba*I and *Not*I are best for linearization, but *Xho*I, *Bam*HI and *Eag*I are also alright. The enzymes have to be able to function in XB. The linear plasmids are purified with Pharmacia GFX PCR DNA and Gel Purification Kit.
18. Agarose-coated dish: pour 1.5% agarose in 1X MMR into 60-mm petri dishes until just over half full. Before the agarose solidifies, place a small weight boat of about 1 square inch in size (or other item that will leave a similar mould or

depression) on the agarose, so that as the agarose solidifies, a square depression in the agarose remains. The depression will accommodate approx 500 eggs. Wrap the dishes in parafilm and store at 4°C until use.

19. Infusion pump and tubing: an infusion pump (Harvard Apparatus, cat. no. 55-1111) is used with a 5-cc syringe onto which is attached a Tygon tubing (1/32 "ID 3/32" OD cat. no. 14-169-1A). The apparatus should be running continuously at 0.1 mL/h throughout the day of the experiment. The system should be filled with (embryo friendly) mineral oil (Sigma M8410) and air bubbles should be avoided, especially when the loaded needle is inserted into the open end of the tube.

20. Injection needles: a variety of glasses can be used to make injection needles. An important consideration is the thickness of the glass, such that when it is pulled, the opening in the end of the needle will be large in comparison to the thickness of the glass to allow easy passage of sperm out of the needle. We use Drummond 50-µL microcaps (cat. no. 501-000-0500) pulled to an outer diameter of 70–85 µm. As a long gently sloping end with a very sharp tip, and an inner diameter of about 50 µm is required, we clip the ends of the pulled glass with forceps. Though the breaks are random, often a tip of the desired shape and size is attained. We do not coat or wash our needles before use.

3. Methods

3.1. Analysis of TR Function During Development Through mRNA Injection into Fertilized Eggs

In the 1980's, Krieg and Melton (1984 and 1987) expanded the use of the frog as a developmental model when they demonstrated that *Xenopus* oocytes and embryos could translate microinjected in vitro transcribed mRNA *(37,38)*. Thus, one could not only use the frog oocyte and embryo as a chemical store, with which one could translate and isolate a protein of interest, but also use the injected mRNA as a tool to examine a protein's function during early development.

The procedure requires the use of in vitro transcribed mRNA that has all the features necessary for translation (5' cap, Kozak sequences, stable 5' and 3' untranslated regions (UTRs), poly(A) tail, etc.). These features are easily introduced into any gene of interest with the use of a good transcription plasmid that has been designed for producing *Xenopus* transcripts. The classical plasmid used, pSP64T *(38)* contained 5' and 3' *Xenopus* β-globin UTRs, into which the coding region of the gene of interest could be inserted. The plasmid also provided simple cloning sites for insert cloning, as well as the sequences needed for polyadenylation (added during in vitro transcription) and proper translation in *Xenopus*. There are now many variations of this plasmid that allow for easier cloning and transcription with a variety of RNA polymerases. A useful site for such plasmids and other information is the "*Xenopus* Molecular Marker Resource" (http://vize222.zo.utexas.edu/).

The following are the procedures that we used to analyze TR/RXR function during frog development *(8,40)*.

3.1.1. In Vitro Transcription

1. Linearize plasmid DNA with an appropriate restriction enzyme, such that the RNA polymerase transcription with produce senses mRNA suitable for transcription (*see* **Note 3**). Prepare enough linearized DNA (e.g., digest about 20 µg) for at least several transcription reactions at 1 µg per reaction.
2. Purify DNA, quantitate, and confirm linearization, purity, and quantification by running a small aliquot on an agarose gel (*see* **Note 4**).
3. Transcribe RNA following directions of Ambion's mMessage mMachine kit (*see* **Note 5**). In our studies TRα and RXRα plasmids *(26)* were linearized with *Eco*RI, twice phenol–chloroform extracted, and ethanol precipitated. One microgram of the DNA was transcribed in vitro with SP6 RNA polymerase.
4. Remove template DNA, purify RNA, and quantitate it (*see* **Note 6**). RNA can be aliquoted and stored at –80°C for months.

3.1.2. In Vitro Fertilization of Xenopus Eggs

1. *X. laevis* females are primed with 75 U of PMSG 3–7 d prior to ovulation.
2. The primed frogs are then injected with 400–600 U (depending on the size of the frog) of HCG the evening prior to the morning that ovulation is desired.
3. A *Xenopus* male frog is sacrificed, the testes are removed, and can be stored in 1X MMR at 4°C for up to 1 wk.
4. Eggs are expelled from ovulating females with gentle squeezing onto petri dishes containing 1 mL 1X MMR into which a small fragment of testes has recently been macerated.
5. After gentle splaying out of eggs into the sperm solution and 2–4 min incubation, the dishes are flooded with 0.1X MMR and incubated for an additional 15 min.
6. To remove the jelly coat, the fertilized eggs are washed with 2.5% cysteine, pH 8.0, until they started to touch each other (about 3–5 min), and then washed 4 to 6 times with 0.1X MMR before being transferred into fresh 0.1X MMR for rearing (*see* **Note 7**).

3.1.3. Embryo Injection and Culturing

1. Healthy-looking embryos are collected just after the beginning of the first division (about 90 min following fertilization) and immediately transferred to 0.5X MMR with 2% Ficoll.
2. After 5 min of incubation in this medium, embryos are injected on both sides with RNA (for TR/RXR a total of 0, 5, 50, or 500 pg of each was used) in a 5-nL total vol per embryo (*see* **Note 8**).
3. Control and injected embryos are kept in 0.5X MMR with 2% Ficoll for 4 to 6 h after injection and then transferred to 0.1X MMR for rearing.
4. Embryos are incubated without any hormone or in the presence of 10–100 n*M* T3. T3 is added to the medium immediately after injection and is present throughout the embryo culturing. The culture medium is changed daily.

5. Following 24, 48 h of growth (or other desired times) embryos are phenotypically examined and sorted (*see* **Note 9**).
6. Desired embryos are selected for Northern blot or other analysis (*see* **Note 9**).

3.2. Use of Chromatin Immunoprecipitation for Investigating TR Action in Development

Anuran metamorphosis is controlled by thyroid hormones (T3). T3 exerts its effects on target tissues via binding to TRs. The presence of TRs in premetamorphic as well as metamorphosing tadpoles, but not embryos, suggests several testable hypotheses regarding TR binding to its target sites in development and chromatin remodeling including histone acetylation as reviewed in the **Subheading 1.**

The occupancy of DNA binding sites for transcription factors, the recruitment of cofactors by transcription factors, as well as histone modifications, such as acetylation, could be addressed using the ChIP method. This technique can be applied to the amphibian model on chromatin isolated from whole animals or organs, depending on developmental stages or T3 status. Indeed, we have successfully used it to follow binding of TR to DNA in vivo, the recruitment of histone deacetylases, and the level of histone H4 acetylation at T3 target genes *(32)*.

3.2.1. Tadpoles Treatment and Tissue Preparation

1. Animals are treated with 10 or 100 n*M* T3 and/or 100 n*M* TSA, a specific histone deacetylase inhibitor.
   For each experiment the number of animals used was 100 embryo at stage 20, 20 tadpoles at stage 47, 1 tadpoles at stage 55 for whole animal analysis, or 10 for tissue-specific analysis.
2. Animals are sacrificed by decapitation after anesthesia by placing them on ice for 20 min. For isolation of nuclei (below), embryos, whole tadpoles at stage 47 or 55, or isolated tails of tadpoles at stage 55, are placed in nuclei extraction buffer A. To isolate nuclei from the intestine, the anterior part of the intestine is dissected and places it in 68% L15 (Leibovitz 15) medium. Clean the interior part of the intestine by flushing it with a needle containing the same medium.
3. Using a homogenizer at very low speed, homogenized whole animals or isolated tail in nuclei extraction buffer A (1 mL of buffer per tadpole or 10 tails). Transfer the homogenate into a dounce to finish homogenization.
4. Homogenize embryos or intestine with a dounce in nuclei extraction buffer A (1 mL of buffer per 100 embryos or 10 intestines).

3.2.2. Nuclei Extraction

1. Spin the homogenate in a swinging bucket at 130,000*g* at 4°C for 3 h.
2. Remove the supernatant and resuspend the pellet in 1 mL of nuclei extraction buffer B.
3. Spin for 5 min in an Eppendorf at 6000*g* at 4°C.

4. Remove the supernatant and resuspend the pellet in 360 μL of nuclei extraction buffer B.

3.2.3. Chromatin Immunoprecipitation

1. Crosslink protein to DNA by adding 10 μL of 37% formaldehyde (1% final, *see* **Note 2**) directly to the nuclei extract and incubate on ice for 10 min and then at room temperature for 20 min.
2. Spin for 5 min at 6000*g*. Remove the supernatant and resuspend the pellet in 200 μL of lysis buffer. Incubate the mixture on ice for 10 min.
3. Sonicate the lysate to reduce DNA length to between 200 and 1000 bp (*see* **Note 10**). Keep the samples on ice all the time.
4. Remove debris by centrifugation for 10 min at 15,000*g* at 4°C.
5. Quantify the amount of DNA in the supernatant by measuring absorption at 260 nm. Dilute each sample to 0.1 μg/ μL in lysis buffer.
6. Take 200 μL of the diluted supernatant and dilute it 10-fold in ChIP dilution buffer.
7. Save 1% of this chromatin solution as the input control.
8. To reduce nonspecific background, preclear the chromatin solution with 80 μL of salmon sperm DNA/protein A agarose slurry for 30 min at 4°C with agitation.
9. Pellet beads by centrifugation for 3 min at 1000*g* at 4°C and collect supernatant.
10. Add 5 to 8 μL of an antibody (against the protein of interest) to 1 mL of the chromatin solution and incubate overnight at 4°C with rotation (10 rpm). The remaining 1 mL of the chromatin solution will be used as no-antibody control.
11. Precipitate the antibody–protein complexes by adding 60 μL of salmon sperm DNA/Protein A Agarose slurry and mixing by rotation at 4°C.
12. At this step, it is time to prepare elution buffer.
13. Pellet immunoprecipitate (agarose beads) by centrifugation (3 min at 1000*g* at 4°C).
14. Wash the beads for 5 min with rotation using 1 mL of each of the buffer listed below:
    Low salt immune complex wash buffer.
    High salt immune complex wash buffer.
    LiCl immune complex wash buffer, and twice with TE.
15. Add 250 μL elution buffer to the pelleted beads. Vortex mix briefly and incubate at room temperature for 15 min with rotation. Centrifuge for 3 min at 1000*g* at room temperature and carefully transfer the supernatant to a new tube. Repeat the entire elution procedure and combine the eluates.
16. After the addition of 20 μL of 5 *M* NaCl, reverse the crosslinks by incubating at 65°C for 4 h. Also, do not forget to do the same for the input control sample.
17. Add 10 μL of 0.5 *M* EDTA, 20 μL of 1 *M* Tris-HCl, pH 6.5, and 2 μL of 10 mg/mL proteinase K and incubate for 1 h at 45°C to degrade proteins.
18. The solution is extracted with phenol–chloroform and precipitated with ethanol to recover DNA (*see* **Note 11**). Wash pellets with 70% ethanol and dry it in a speedvac.
19. Resuspend the DNA pellet in 20 μL of water.
20. Detect specific sequences from no-antibody control, immunoprecipitated, input control, and unbound DNA samples by PCR. Conditions for PCR must be determined for each gene of interest. The PCR product size should be between

200 and 400 bp. Addition of labeled ^{32}p-dCTP (1 µCi) allows one to detect the products with high sensitivity by autoradiography.

21. Add 2 µL of 10X DNA loading buffer to 10 µL of the PCR product and load on a 6% non denaturing polyacrylamide Tris-borate EDTA (TBE) gel. After electrophoresis, dry the gel before autoradiography (Ethidium bromide staining can be used as a semiquantitative assay if no radioactive label is used in the PCR).

3.3. Transgenic Analysis of TR Function in Development

Xenopus transgenesis is relatively new but relatively simple and powerful technique for studying gene function in development. As the transgene is integrated into the male genome prior to fertilization, the resulting embryo is not chimeric, and breeding of animals is not required. Many transgenic animals can be generated in a single day and analyzed on the next day or later with relatively little financial cost.

It is a technique with several steps, each fraught with some problems where, in the end, one tries to minimize damages to sperm, which will lead to developmental anomalies, while at same time inducing sufficient cuts/damage into sperm DNA with restriction enzymes in the presence of egg extract to allow incorporation of foreign DNA into the sperm genome. One needs good quality sperm nuclei and egg extract. The optimal use and ratio of these two critical components often have to be determined empirically from batch to batch.

The procedure described here is based on that of Kroll and Amaya (1996) *(35)* with modifications from *(36)* (**Fig. 5**). Additional information is available at the following web pages: (http://www.biosci.utexas.edu/MCDB Xenbase/genetics/transgen.html), (http://www.welc.cam.ac.uk/~ea3/The.Amaya.Lab.Homepage.html), (http://www.virginia.edu/~develbio/trop/overview/transgenesis.html), and (http://www.acs.ucalgary.ca/~browder/frogsrus.html).

3.3.1. Sperm Nuclei Preparation

1. Remove testes and wash once in cold 1X MMR and twice in cold 1X NPB. Clean with fine forceps and scissors to remove blood vessels, fat, and connective tissues, as much as possible without puncturing the testes. Macerate testes (best done with a pair of fine forceps) in a clean dry dish, leaving no obvious large clumps. Do not allow testes to dry.
   Resuspend in a total of 8 mL 1X NPB by gentle pipeting up and down with fire polished pasteur pipet (3-mm opening) or similar large bore pipet (always use a large bore pipet and gentle force). Filter suspension through cheesecloth into 15 mL tube (Falcon 2059) squeezing extra residue from cheesecloth with a gloved hand (*see* **Note 12**).
2. Spin the filtrate once at 1483*g* (3000 rpm) for 15 min at 4°C in a Sorvall HB-4 rotor. All spins are performed in Falcon 2059 tubes.
3. Set up a step gradient: 5 mL 50% Percoll (Sigma, cat. no. P-1644) in 1X NPB at the bottom and , 5 mL 25% Percoll in 1X NPB at the top (*see* **Note 13**).

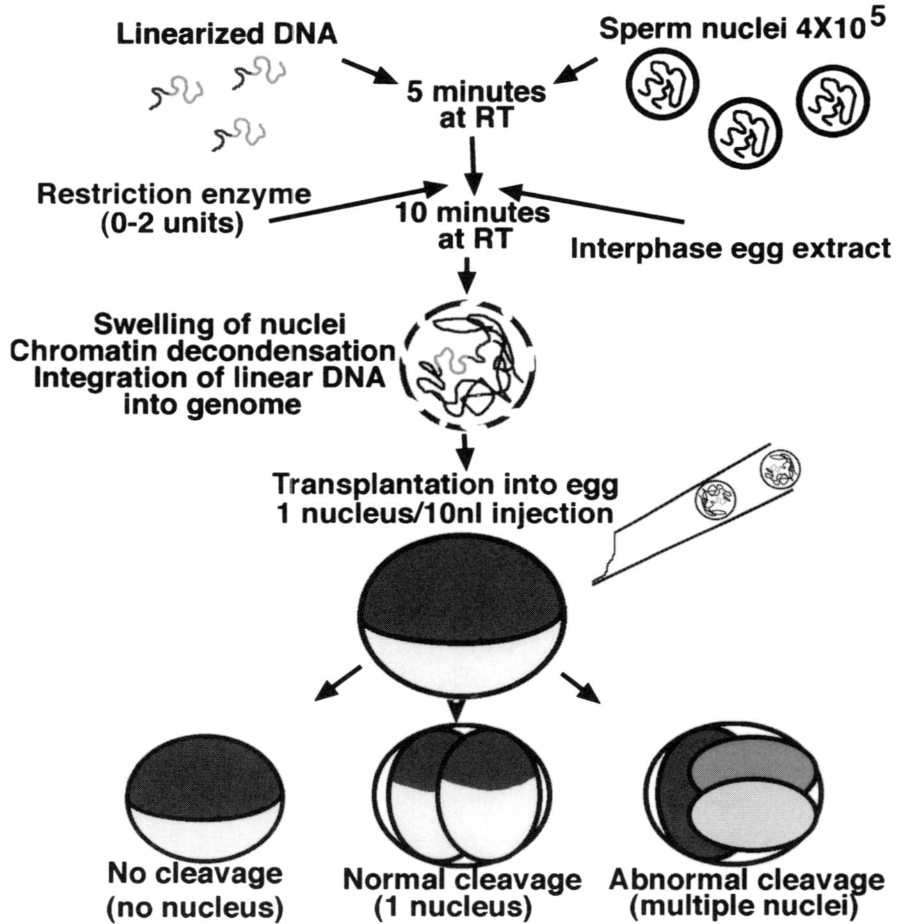

Fig. 5. Schematic flow chart of the transgenic procedure of Kroll and Amaya (1996) *(35)*.

4. Carefully resuspend sperm pellet in 3 mL 1X NPB, leaving red blood cells and other visible nonsperm cells in the pellet if possible.
5. Layer sperm on top of Percoll gradient and spin at 1483*g* (3000 rpm) for 15 min at 4°C in a Sorvall HB-4 rotor.
   Mature sperm should be the bottom (densest) pellet (to make sure, quickly check other layers with a microscope).
6. Remove as much of Percoll as possible and carefully resuspend mature sperm in cold 8 mL 1X NPB.
7. Spin the resuspension at 1483*g* (3000 rpm) for 15 min at 4°C in a Sorvall HB-4 rotor and carefully resuspend sperm in 1 mL 1X NPB (to wash out residual Percoll).

8. Add 5 µL of 10 mg/mL digitonin to the sperm resuspension and mix gently (*see* **Note 14**), then incubate the mixture at room temperature for 5 min.
9. Add 10 mL cold NPB with 3% BSA (to stop digitonin action) and spin at 1483*g* (3000 rpm) for 15 min at 4°C in a Sorvall HB-4 rotor.
10. Carefully resuspend the pellet in 8 mL cold NPB (to wash out excess BSA) and spin at 1483*g* (3000 rpm) for 15 min at 4°C in a Sorvall HB-4 rotor.
11. Carefully resuspend the pellet in 500 µL NPB with 30% glycerol and 0.3% BSA. The sperm nuclei can be used fresh. They can also be stored overnight at 4°C to allow optimal penetration of glycerol into nuclei, then aliquoted, flash-frozen in liquid nitrogen and stored at –80°C for several months (*see* **Note 15**).

3.3.2. High Speed Egg Extract Preparation

All reagents should be prepared beforehand, and the procedure should be carried out promptly once initiated. Optimally, the high-speed spin should begin within 45–60 min of dejellying the eggs.

Prepriming female frogs with PMSG is not needed if you have proven good ovulators.

1. Prime 6–10 females with 600–700 U of HCG and leave them overnight in separate buckets in about 2 to 3 L of 1X MMR.
2. Collect eggs the following morning from the 1X MMR. Do not use eggs from a particular frog if any of the eggs are lysing or mottled. Can expel more eggs into the 1X MMR by squeezing the frog (if needed).
3. Dejelly eggs in 2.5% cysteine in 1X MMR (do not dejelly for longer than 5 to 6 min as this will damage eggs). Wash eggs 4× in 1X MMR, 35 mL. Remove debris, lysed, bad, and activated eggs as best you can. This may take a while (*see* **Note 16**)
4. Wash eggs with 2× 25 mL CSF-XB (*see* **Note 17**) and transfer them to 14 × 95 mm tubes (Falcon 2059) (to remove excess CSF-XB).
5. Add 1 mL Versilube F-50 to the top of the eggs (*see* **Note 18**) and spin at room temperature for 1 min at the (1000 rpm) 150*g* and then 30 s at (2000 rpm) 600*g* in a clinical tabletop centrifuge. The eggs should be packed but not broken.
6. Remove excess CSF-XB and balance tubes, and then spin 10 min at 16,500*g* (10,000 rpm) at 4°C in Sorvall HB4 rotor. Three layers should form: lipid (top), cytoplasm, and yolk (bottom).
7. Remove the cytoplasm through the side of the tube with an 18-gauge needle. The color should be relatively golden. If it is predominantly gray, respin.
8. Add 1/20th vol ATP energy mixture and transfer the clarified cytoplasm into Beckman TL100.3 tubes.
9. Add $CaCl_2$ to a final concentration of 0.4 m*M* and incubate for 15 min at room temperature (this inactivates CSF and pushes the extract into interphase).
10. Balance the tubes and spin at 259,000*g* (55,000 rpm) for 1.5 h at 4°C (Beckman TLS 55 rotor, TLX ultracentrifuge). The resulting layers, from top to bottom should be lipid, cytosol, membranes/mitochondria, and glycogen/ribosomes.

11. Carefully aspirate off lipids, and remove cytosol and respin the cytosol at 259,000g (55,000 rpm) for 30 min at 4°C. The cytosol should be clear.

12. Aliquot the clear cytosol, quick-freeze it with liquid NO_2, and store the aliquots at –80°C (*see* **Note 19**).

3.3.3. Restriction Enzyme-Mediated Integration (REMI) Reaction

1. Inject the desired number of females (2–4) with 600–700 U of HCG (subcutaneously in the hind leg) the night prior to the experiment (so they will ovulate the following morning).

2. If reagents and materials were stored in the cold, make sure they reach room temperature before use.

3. Fill agarose well in injection dish with 0.6X MMR and 6% Ficoll.

4. Thaw sperm nuclei and egg extract on ice. Do not use after about 90 min on ice (*see* **Note 20**). Thaw additional samples as needed. Do not refreeze them. Also, thaw sdb and DNA. Make a 1:20 dilution of the desired restriction enzyme (usually the same one as was used to linearize the plasmid) in water, and store on ice (do not use after 60 min) (*see* **Note 21**).

5. Gently mix 4 µL sperm nuclei stock (total of 250,000–300,000 sperm), 5 µL linear plasmid (150–250ng/µL), and sdb added to 15 µL, and incubate the mixture for 5 min room temperature (*see* **Note 22**).

6. Add 1 µL of 1:20 dilute restriction enzyme (usually a 6 bp cutter, such as *Eag*I and *Xba*I, but it also works with a 8 bp cutter such as *Not*I) and 5–15 µL high-speed extract (depending on quality) to the above mixture and mix gently with a cut yellow tip (used for 200-µL pipetman). Incubate the resulting sample for 10 min at room temperature (sperm should swell).

7. Dilute the reaction mixture with sdb to about 10 sperm/nL (*see* **Note 23**).

8. Collect eggs from female and dejelly them with 2.5% cysteine in 1X MMR (should be done during the 5- and 10-min incubations in this protocol). Select healthy unactivated eggs, and transfer them to the agarose well in 0.6X MMR and 6% Ficoll (*see* **Note 24**). After 5 min in 0.6X MMR and 6% Ficoll, the eggs will easily pierce.

9. Load sperm reaction mixture into an injection needle, from back, using a yellow pipet tip onto which a short (5 mm) piece of Tygon tubing has been attached (*see* **Note 25**).

10. Insert loaded needle into a tubing that is attached to the infusion pump (insert needle gently). Note that tubing becomes brittle with time and will need to be replaced when needle insertion becomes difficult. The needle on the tubing is then placed on an electrode or any other type of needle holder mounted on the micromanipulator, which will be used for the injections.

11. Inject perpendicular to egg surface membrane (if needle is not sharp, it will cause much damage). Pierce the egg with quick inward jabs, and remove the needle more slowly. Develop a rhythm that is quick, but results in minimum damage and cytoplasm leakage out of the wound left in the egg. Continuously watch to see that the needle does not become clogged as sperm nuclei will be damaged or blocked, leading to the formation of haploid embryos.

12. About 15 min after injection, gently remove the injected eggs from the agarose well and place then in a fresh plate in 0.6X MMR and 6% Ficoll. This will allow them better oxygen exchange and will provide for easier embryo sorting later (*see* **Note 26**).
13. Prior to gastrulation (about stage 6 or 7) transfer the healthy embryos into 0.1X MMR and 6% Ficoll, with 50 µg/mL gentamicin. Following gastrulation, embryos can be transferred into 0.1X MMR and reared normally (*see* **Notes 26** and **27**).

4. Notes

1. A variety of instruments and approaches can be used to pull needles and inject RNA or DNA samples. They include simple hand pulled needles using a Bunsen Burner, and forced air injectors using "house air" found in most laboratories. The crucial criteria are choosing a needle and a method that allow for a small needle diameter (about 10 µm) to be pulled and a way to calibrate or control injection volumes (about 5–30 nL/injection).
2. Formaldehyde should be used in a fume hood. All chemicals should be considered potentially hazardous and handled with care.
3. When linearizing plasmids, avoid excess DNA sequences following the gene of interest (such as long multiple cloning sites found in some plasmids). If these are transcribed, they may form RNA secondary structures that may inhibit efficient translation. In addition, avoid linearizing DNA with restriction enzymes which leave 3' overhangs (such as *PST*I), as that may prime transcription from the inappropriate strand *(41)*.
4. Use gel electrophoresis to confirm that the DNA is pure and completely linearized. Circular plasmid templates will generate extremely long heterogeneous RNA transcripts, because RNA polymerases are very processive.
5. Though a commercial RNA transcription kit is optimized for producing translatable RNA, cost may prohibit the use of such a kit. Transcription can easily be performed with transcription components being purchased individually. RNA polymerases and their appropriate buffer, nucleotides, m7G(5')ppp(5')G cap analogs, DNase, and other reagents are available from numerous companies. Care must be taken, however, to ensure that when these reagents are used together, conditions are optimized for efficient transcription. This may take some trial and error.
6. Once the transcription reaction is complete (following the removal of the DNA template with DNase) the RNA must be purified prior to its injection. LiCl precipitation is a good and quick method of removing unincorporated nucleotides and proteins. Phenol–chloroform extraction, followed by ehtanol precipitation, is a more stringent method of purification, though care must be taken to remove all traces of phenol and chloroform prior to injection.
7. For a detailed look at *Xenopus* fertilization protocols (*see* **ref.** *42*).
8. Once injected, RNA stability (half-life) is dependent on a number of factors including transcript size, secondary structure, 5' and 3' UTR lengths and sequences, poly(A) tail lengths, and others. Most injected RNA (such as TR or

RXR) can still be readily detectable after 10 h, though Northern or other analysis should be used to determine the half-life of the transcript used if needed. Injecting higher doses of RNA will not aid in extending the availability of RNA as injected RNA (and DNA) is toxic to the embryo at high doses. Standard nucleic acid toxicity phenotypes include gastrulation defects (ring embryos, sometimes called the spina bifida phenotype), kinked embryos (abnormal somitogenesis), anterio-dorsal deficiencies (small heads or eyes, lack of pigment cells), and death. Nucleic acids also become toxic at much lower concentrations if too great a volume is injected *(41)*.

9. Following injection and initial sorting of embryos, analysis and phenotypic classification depends on the function of the gene of interest. TR/RXR overexpression produces distinct phenotypes in the absence and presence of T3. Scoring (classification) of embryos much be achieved with readily visible criteria. Thus, axis length, body width, presence/size of eye or cement gland, and other finer details such as number and organization of somites can all be used to correlate a treatment with a given number of defects. Subsequently, more detailed molecular or biochemical analyses can be carried out. RNA can be isolated from embryos and Northern blot analysis used to examine changes of expression in downstream genes. Soluble proteins can be isolated from embryos and used in Western analysis of various proteins of interest, or in gel mobility shift assays examining binding of overexpressed TR/RXR to TREs. Finally, whole mount *in situ* hybridization analysis can be used to investigate gene expression, etc.

10. To establish optimal conditions required to shear crosslinked DNA to 200–1000 bp in length, a mock experiment should be done where the number of 10-s pulses and/or the power setting is varied. Our experience shows that DNA is sheared to the appropriate length with 15 sets of 10-s pulses using a sonicator equipped with a 2-mm tip and set to duty cycle just under constant and output on 7 (Branson Sonifier 450, VWR Scientific). Ear protection should be worn during sonication.

11. Addition of an inert carrier, such as 20 µg of glycogen or yeast RNA, is suggested to facilitate the recovery of DNA.

12. Protease inhibitors used in the original protocol during sperm nuclei preparation are not used. The cheesecloth, which is normally used in cheese making during a straining process and compaction of the curds, is used here to strain out large particulate pieces of testis. The exact pore size of the cheesecloth is not critical.

13. There are several ways to separate the healthy mature sperm from other cell types. The percoll step gradient of **ref. *36*** provides a relatively simple way to isolate the dense mature sperm, though several other differential centrifugation methods may be used. Simple monitoring of sperm with microscopy should be carried out to ensure that a final product of mature sperm is as desired.

14. Though several detergents are available and function in the removal of membranes (including lysolecithin and Triton X-100), digitonin was used due to its gentleness and ease of preparation and storage.

15. To quantitative sperm, dilute 1 µL sperm in 100 µL sperm dilution buffer and add 1 µL 1:100 Hoechst. Count on hemacytometer (count sperm in 4 × 4 box, there

are 4 4 × 4 boxes, and the average number of cells in a 4 × 4 box, e.g., 55, gives you 55 × 10^4 cells/mL). Typically, you should have 75–125 sperm/nL. If you have less than 50/nL, repellet and resuspend in smaller volume. Never refreeze sperm nuclei.

16. Good quality eggs are essential for a good quality egg extract, though the presence of several poor eggs in the initial batch may be somewhat tolerated. Although egg extracts prepared for the studies of cell cycle or nuclear assembly must be pure *(43)*, the qualities required here are less stringent. The potency of the extract to decondense sperm nuclei will be reduced in poor preparations, but they are very often well within useable limits.

17. The original protocol *(35)* called for the use of protease inhibitors during the preparation of the egg extract. These inhibitors are essential to producing egg extracts that will function in nuclear assembly and cell cycling studies, but is less essential here. The final test of the quality of the egg extract, for our purposes, is its ability to decondense sperm chromatin.

18. Versilube and other oils such as Nyosil M-25, are used because of their density and inertness. They displace water from among the eggs, resulting in a more concentrated and potent egg extract, without effecting its function. Their use may increase the potency of the extract, but they are not essential.

19. If egg extract is good, Hoechst-stained sperm nuclei should swell very significantly in about 10 min at room temperature when viewed under a microscope.

20. Once sperm nuclei are decondensed they should never be placed back on ice.

21. The use of restriction enzymes with a particular batch of sperm or egg extract may need to be determined empirically. If the sperm nuclei preparation is somewhat damaged, then restriction enzymes may not be needed to facilitate nicking of the DNA. Similarly, the potency of the egg extract affects the time needed to ligate and repair DNA, and the amount of the restriction enzyme that should be used.

22. Damaged sperm is the Achilles tendon of this procedure and so care should be taken during manipulations to decrease damage. Always cut the tip off of narrow micropipet tips (yellow, 200 µL tips used for pipetman) and avoid introducing air bubbles into sperm-containing solutions.

23. As we start with 250,000 initial sperms, we dilute the REMI reaction by adding 40 µL of sdb. We then further dilute by removing 10 µL and adding it to 50 µL fresh sdb, which is then used for injection.

24. Do not spend too much time sorting. If egg quality is poor, discard the entire batch and use eggs from a different female.

25. Place the back of the needle firmly against the tubing to load. There is no need to insert needle into tube (in fact, this is detrimental as it will cause problems due to back-flow when removed from the tubing).

26. The first cleavage is an important gauge of the experimental procedures and reagents. Embryos (20–60%) should have normal 2-cell cleavages. Multiple cleavage furrows suggest multiple nuclei were injected, and therefore, the sperm dilution should be increased or the injection rate decreased. Other strange cleav-

age patterns and marbled pigmentation could result due to high injection volume, or due to toxic reagents. Gastrulation is also an important gauge of the quality of the sperm at the time of injection. Exogastrulation and other incomplete gastrulation phenotypes are possibly due to damaged sperm, which may have occurred at any time during the procedure. In our study, the transgenes are fused to Green fluorescent protein (GFP), and assays for GFP expression can be carried out at any time after midblastula transition. Lack of normal embryos with detectable GFP suggests no incorporation of plasmid DNA into the sperm, which could be indicative of incomplete removal of the sperm membrane, or poor quality egg extract that does not decondense the sperm chromatin and allow DNA integration.

27. As there are many GFP varieties and extensive autofluorescence in many tissues in the frog, it is essential that a good GFP filter sets be used. Chroma's GFP II filter set is a good all-round solution.

Acknowledgments

We would like to thank Dr. K. L. Kroll for advice and help in the transgenesis procedure. S. Damjanovski and L. M. Sachs are equal contributors.

References

1. Shi, Y.-B. (1999) *Amphibian Metamorphosis: From Morphology to Molecular Biology.* John Wiley and Sons, Inc., New York, p. 288.
2. Tata, J. R. (1993) Gene expression during metamorphosis: an ideal model for post-embryonic development. *Bioessays* **15,** 239–248.
3. Hetzel, B. S. (1989) *The Story of Iodine Deficiency: An International Challenge in Nutrition.* Oxford University Press, Oxford.
4. Dodd, M. H. I. and Dodd, J. M. (1976) The biology of metamorphosis, in *Physiology of the Amphibia* (Lofts, B., ed.) Acadmic Press, New York, pp. 467–599.
5. Lazar, M. A. (1993) Thyroid hormone receptors: multiple forms, multiple possibilities. *Endocr. Rev* **14,** 184–193.
6. Mangelsdorf, D. J. and Evans, R. M. (1995) The RXR heterodimers and orphan receptors. *Cell* **83,** 841–350.
7. Tsai, M. J. and O'Malley, B. W. (1994) Molecular mechanisms of action of steroid/thyroid receptor superfamily members. *Ann. Rev. Biochem* . **63,** 451–486.
8. Puzianowsak-Kuznicka, M., Damjanovski, S., and Shi, Y.-B. (1997) Both thyroid hormone and 9-cis retinoic acid receptors are required to efficiently mediate the effects of thyroid Hormone on embryonic development and specific gene regulation in *Xenopus laevis. Mol. Cell Biol.* **17,** 4738–4749.
9. Shi, Y.-B., Wong, J., and Puzianowska-Kuznicka, M. (1996a) Thyroid hormone receptors: Mechanisms of transcriptional regulation and roles during frog development. *J. Biomed. Sci.* **3,** 307–318.
10. Perlman, A. J., Stanley, F., and Samuels, H. H. (1982) Thyroid hormone nuclear receptor. Evidence for multimeric organization in chromatin. *J. Biol. Chem.* **257,** 930–938.
11. Wong, J., Shi, Y. B., and Wolffe, A. P. (1995) A role for nucleosome assembly in

both silencing and activation of the Xenopus TR beta A gene by the thyroid hormone receptor. *Genes Dev.* **9,** 2696–2711.

12. Wolffe, A. P. (1997) Chromatin remodeling regulated by steroid and nuclear receptors. *Cell Res.* **7,** 127–142.
13. Burke, L. J. and Baniahmad, A. (2000) Co-repressors 2000 [In Process Citation]. *FASEB J.* **14,** 1876–1888.
14. Chen, J. D. and Li, H. (1998) Coactivation and corepresion in transcriptional regulation by steroid/nuclear hormone receptors. *Crit. Rev. Eukaryo. Gene Expr.* **8,** 169–190.
15. Guenther, M. G., Lane, W. S., Fischle, W., Verdin, E., Lazar, M. A. and Shiekhattar, R. (2000) A core SMRT corepressor complex containing HDAC3 and TBL1, a WD40-repeat protein linked to deafness. *Genes Devel.* **14,** 1048–1057.
16. Hu, X. and Lazar, M. A. (2000) Transcriptional repression by nuclear hormone receptors. *TEM* **11,** 6–10.
17. Li, J., Wang, J., Wang, J., et al. (2000) Both corepressor proteins SMRT and N-CoR exist in large protein complexes containing HDAC3. *EMBO J.* **19,** 4342–4350.
18. McKenna, N. J., Lanz, R. B., and O'Malley, B. W. (1999) Nuclear receptor coregulators: cellular and molecular biology. *Endocr. Rev.* **20,** 321–344.
19. Urnov, F. D., Yee, J., Collingwood, T. N., et al. (2000) Targeting of N-CoR-HDAC3 by the oncoprotein v-ErbA yields a chromatin infrastructure- dependent transcriptional repression pathway. *EMBO J.* **19,** 4074–4090.
20. Xu, L., Glass, C. K., and Rosenfeld, M. G. (1999) Coactivator and corepressor complexes in nuclear receptor function. *Curr. Opin. Genet. Dev.* **9,** 140–147.
21. Hsia, S.-C. V., Wang, H., and Shi, Y.-B. (2001) Involvement of Chromatin and Histone Acetylation in the Regulation of HIV-LTR by Thyroid Hormone Receptor. *Cell Res.*, **11,** 8–16.
22. Wong, J., Patterton, D., Imhof, D., Guschin, D., Shi, Y.-B., and Wolffe, A. P. (1998) Distinct requirements for chromatin assembly in transcriptional repression by thyroid hormone receptor and histone deacetylase. *EMBO J.* **17,** 520–534.
23. Rachez, C. and Freedman, L. P. (2000) Mechanisms of gene regulation by vitamin D(3) receptor: a network of coactivator interactions. *Gene* **246,** 9–21.
24. Sachs, L. M., Damjanovski, S., Jones, P. L., et al. (2000) Dual functions of thyroid hormone receptors during Xenopus development. *Comp. Biochem. Physiol. B. Biochem. Mol. Biol.* **126,** 199–211.
25. Yaoita, Y., Shi, Y. B., and Brown, D. D. (1990) Xenopus laevis alpha and beta thyroid hormone receptors. *Proc. Natl. Acad. Sci. USA* **87,** 7090–7094.
26. Wong, J. and Shi, Y.-B. (1995) Coordinated regulation of and transcriptional activation by Xenopus thyroid hormone and retinoid X receptors. *J. Biol. Chem.* **270,** 18,479–18,483.
27. Shi, Y.-B., Wong, J., Puzianowska-Kuznicka, M., and Stolow, M. (1996b) Tadpole competence and tissue-specific temporal regulation of amphibian metamorphosis: roles of thyroid hormone and its receptors. *Bioessays* **18,** 391–399.
28. Machuca, I., Esslemont, G., Fairclough, L. and Tata, J. R. (1995) Analysis of structure and expression of the Xenopus thyroid hormone receptor b gene to explain its autoregulation. *Mol. Endocrinol.* **9,** 96–107.

29. Ranjan, M., Wong, J., and Shi, Y. B. (1994) Transcriptional repression of Xenopus TR beta gene is mediated by a thyroid hormone response element located near the start site. *J. Biol. Chem.* **269,** 24,699–24,705.

30. Shi, Y.-B. (1996) Thyroid hormone-regulated early and late genes during amphibian metamorphosis, in *Metamorphosis: Post-Embryonic Reprogramming of Gene Expression in Amphibian and Insect Cells.* (Gilbert, L. I., Tata, J. R., and Atkinson, B. G., eds.) Academic Press, New York, pp. 505–538.

31. Furlow, J. D. and Brown, D. D. (1999) In vitro and in vivo analysis of the regulation of a transcription factor gene by thyroid hormone during *Xenopus laevis* metamorphosis. *Mol Endocrinol.* **13,** 2076–2089.

32. Sachs, L. M. and Shi, Y.-B. (2000) Targeted chromatin binding and histone acetylation in vivo by thyroid hormone receptor during amphibian development. *Proc. Natl. Acad. Sci. USA* **97,** 13,138–13,143.

33. Yaoita, Y. and Brown, D. D. (1990) A correlation of thyroid hormone receptor gene expression with amphibian metamorphosis. *Genes Dev.* **4,** 1917–1924.

34. Ishizuya-Oka, A., Ueda, S. and Shi, Y. B. (1997) Temporal and spatial regulation of a putative transcriptional repressor implicates it as playing a role in thyroid hormone-dependent organ transformation. *Dev. Genet.* **20,** 329–337.

35. Kroll, K. L. and Amaya, E. (1996) Transgenic *Xenopus* embryos from sperm nuclear transplantations reveal FGF signaling requirements during gastrulation. *Development* **122,** 3173–3183.

36. Huang, H., Marsh-Armstrong, N., and Brown, D. D. (1999) Metamorphosis is inhibited in transgenic Xenopus laevis tadpoles that overexpress type III deiodinase. *Proc. Natl. Acad. Sci. USA* **96,** 962–967.

37. Nieuwkoop, P. D. and Faber, J. (1956) Normal table of *Xenopus laevis*, 1st ed., North Holland, Amsterdam, The Netherlands.

38. Krieg, P. A. and Melton, D. A. (1984) Functional messenger RNAs are produced by SP6 in vitro transcription of cloned cDNAs. *Nucl. Acids Res.* **12,** 7057–7070.

39. Krieg, P. A. and Melton, D. A. (1987) In vitro RNA synthesis with SP6 RNA polymerase. *Methods Enzymol.* **155,** 397–415.

40. Damjanovski, S., Puzianowska-Kuznicka, M., Ishuzuya-Oka, A. and Shi, Y. B. (2000) Differential regulation of three thyroid hormone-responsive matrix metalloproteinase genes implicates distinct functions during frog embryogenesis. *FASEB J.* **14,** 503–510.

41. Vize, P. D., Melton, D. A., hemmati-Brivanlou, A., and Harland, R. M. (1991) Assays for gene function in developing *Xenopus* embryos, in *Methods Cell Biol.* **36,** 367–387.

42. Wu, M. and Gerhart, J. (1991) Raising *Xenopus* in the Laboratory. *Meth. Cell Biol.* **36,** 3–18.

43. Murray, A. W. (1991) Cell cycle extracts. *Methods Cell Biol.* **36,** 581–605.

44. Leloup, J. and Buscaglia, M. (1977) La triiodothyronine: hormone de la métamorphose des amphibiens. *C. R. Acad. Sci.* **284,** 2261–2263.

10

Transcriptional Regulation by Thyroid Hormone Receptor in Chromatin

Jiemin Wong

1. Introduction

Genomic DNA in eukaryotic cells is packaged with histone and nonhistone proteins into chromatin structure. Both biochemical and genetic evidences indicate that chromatin structure imposes constraints on nuclear processes including transcription, replication, recombination and repair *(1,2)*. Thus, a central question in study of transcription regulation is how transcription factors function in the context of chromatin. In recent years, it has become increasingly clear that chromatin structure has an important role in regulating gene expression and that transcription factors can actively recruit chromatin-remodeling enzymes to regulate transcription of their target genes positively or negatively *(3–6)*. In this regard, thyroid hormone receptor (TR) is one of the best-studied transcription factors and has contributed its share of information to the current concept of transcription regulation *(7–9)*.

As a member of nuclear hormone receptor superfamily, early studies using transient transfection indicated that TR has the capacity to repress transcription in the absence of T3 and activate in the presence of T3 *(10,11)*. In order to study how TR regulates transcription in chromatin, we have chosen to use *Xenopus* oocyte as a model system. The *Xenopus* oocyte offers a unique system to study transcription owing to its large storage of basal transcription factors and proteins involving in chromatin assembly *(12)*. The *Xenopus* oocytes are well suited for introducing DNA, mRNA or proteins through microinjection and have long been used for transcription analysis. Importantly, the plasmid DNA injected into the nucleus is assembled into chromatin, and the types of chromatin formed differently dependent on the forms of the injected DNA *(13)*. When double-stranded

From: *Methods in Molecular Biology, Vol. 202: Thyroid Hormone Receptors: Methods and Protocols*
Edited by: A. Baniahmad © Humana Press Inc., Totowa, NJ

reporter DNA (dsDNA) is injected, it will be assembled in a relatively slow fashion (5 to 6 h) into chromatin with less defined nucleosomal arrays, which permits a relatively high level of basal transcription. However, when single-stranded reportor DNA (ssDNA) is injected, it will quickly undergo synthesis of the complementry strand and is assembled into chromatin in a replication-coupled pathway, mimicking the genomic DNA chromatin assembly process in somatic cells *(13)*. The resulting chromatin usually yields a lower level of basal transcription. Thus, *Xenopus* oocytes provide a unique opportunity to study the transcriptional regulation under different chromatin condition by using different forms of reporter DNA. In addition, *Xenopus* oocytes are well documented for ability to translate injected mRNAs into proteins *(14)*. Thus, it is possible to virtually express any proteins in *Xenopus* oocytes through microinjection of the corresponding mRNAs synthesized in vitro.

Taking these advantages, we have investigated the mechanism by which TR regulates transcription of its target gene (*Xenopus TRβA* gene promoter) in chromatin. We have demonstrated that, regardless of T3, the expression of TR and retinoid x receptor (RXR) heterodimers (TR/RXR) leads to the constitutive occupation of its target sequence (thyroid hormone response element [TRE]) in chromatin as revealed by DNase I hypersensitive assay as well as polymerase chain reaction (PCR)-based DNase I footprinting *(15,16)*. We have also demonstrated that TR/RXR repress transcription in the absence of T3 and activate transcription in the presence of T3 *(17)*. Importantly, we observed that a robust T3-dependent activation by TR/RXR requires the reporter DNA being assembled into chromatin via replication-coupled chromatin assembly pathway, indicating that TR/RXR have evolved to function efficiently in the context of chromatin and make use of chromatin components for transcriptional activation *(17)*. Further analyses of chromatin structure by micrococcal nuclease (MNase) digestion reveal that addition of T3 to chromatin-bound TR/RXR induces, in addition to transcriptional activation, an extensive disruption of regular nucleosomal arrays *(16)*. The disruption of chromatin structure by liganded TR/RXR can be further confirmed by supercoiling assay, which measures the number of nucleosomes on circular DNA molecules on the basis that each nucleosome is known to introduce one negative supercoiling turn into DNA. In this chapter, I will summarize the protocols we have used for analyses of how TR/RXR functions in the context of chromatin.

1.1. Xenopus *Oocytes*

Xenopus laevis can be raised by laboratory as discussed previously *(18)* or purchased from commercial sources. Due to the long life cycle of *Xenopus laevis*, it is more convenient to purchase mature female frogs from commercial sources and maintain them in laboratory. Since each mature female frog contains about tens of thousand oocytes, each frog can be used upon to 3 to 4 times for obtaining oocytes to cut down cost. To accomplish this, we basically use a protocol as described by Smith et al. to anesthetize the frog and surgically remove just a sufficient amount of ovary from the frog *(19)*. In brief, hypothermia is used for anesthesia of frogs. A frog is packed in an ice bucket with ice-water for about 30–40 min and then placed on a bed of ice for subsequent surgery. A small incision (less than 1 cm) is made in the skin and body wall in the posterior ventral side of the animal. The ovary consisting of several lobes can be exposed gently with forceps and one or more lobes can be snipped off using scissors. The incision in the body wall and skin is then sutured back sequentially, and the frog is left to recover in the bucket with room temperature water. The recovery is quite rapid (several minute), and the frog can be returned to the tank for future use.

1.2. Analysis of Chromatin Structure of Reporter DNA in Xenopus Oocytes

In general, assembly of DNA into chromatin inhibits the binding of transcriptional factors and, in particular, the function of basal transcription machinery. Thus, it is not surprising that many transcription factors including TR/RXR have capacity to exert active remodeling or disruption of chromatin structure surrounding their binding sites. Although the exact molecular mechanism underlining the chromatin remodeling is not entirely clear, it is believed to require the involvement of one or both classes of chromatin remodeling activities: ATPase-dependent chromatin remodeling factors, such as SWI/SNF, RSC, NURF, and CRACK, and histone acetyltransferases, such as CBP/p300, PCAF/GCN5 and SRC1/ACTR (for review, *see* **ref. 4**). The disruption of the chromatin structure can be analyzed by MNase digestion, supercoiling assay, and DNase I hypersensitive site assay.

2. Materials

2.1 Preparation of Xenopus Oocytes

1. 10X OR2 buffer: Dissolve to final 1X concentration: 48.2 g NaCl to 82.5 mM, 1.86 g KCl to 2.5 mM, 2.03 g MgCl$_2$ to 1 mM, 1.42 g Na$_2$HPO$_4$ to 1 mM, 11.9 g HEPES to 5 mM.
   Adjust pH to 7.8 with 1 N NaOH. Bring final volume to 1 L with distilled water.
2. 10X MBSH buffer: Dissolve first to final 1X concentration: 23.8 g HEPES to 10 mM, 51.3 g NaCl to 88 mM, 0.75 g KCl to 1 mM, 2 g NaHCO$_3$ to 2.4 mM. Then add to final 1X concentration: 2 g MgSO$_4$·7H$_2$O to 0.82 mM, 0.9 g CaCl$_2$·6H$_2$O to 0.41 mM, 0.78 g Ca(NO$_3$)2·4H$_2$O to 0.33 mM.
   Adjust to pH 7.6 with 1 N NaOH. Bring final volume up to 1 L with water. Add antibiotics penicillin/streptomycin (pen/strep) to a final concentration of 100 µg/mL.

2.2. Microinjection of Xenopus Oocytes

1. Stage VI *Xenopus* oocytes (**Note 1**).
2. Nanoject II from Drummond Scientific Company (**Note 2**).
3. mMESSAGE mMACHINE kits from Ambion (**Note 3**).
4. Nitex mesh.
5. Forceps.
6. Microscope.
7. Injection needle.
8. M13 or phagemid-based vectors such as pBluescript II (**Note 4**).

2.3. Analysis of Transcription From Reporter DNA

1. Oocyte homogenization buffer: 0.1 M Tris-HCl, pH 8.0, 10 mM EDTA.
2. DNA recovery buffer: 0.2 % sodium dodecyl sulfate (SDS), 1 mM EDTA, 10 mM Tris-HCl, pH 8.0.
3. 5X primer extension annealing buffer: 0.1 M Tris-HCl, pH 8.3, 2 M KCl.
4. 10X reverse transcription (RT) buffer: 0.5 M Tris-HCl, pH 8.3, 60 mM MgCl$_2$.
5. Formamide loading buffer: 80 % formamide, 10 mM EDTA, mg/mL xylene cyanol, 1 mg/mL bromophenol blue.

2.4. Analysis of Chromatin Structure of Reporter DNA in Xenopus Oocytes

1. MNase buffer: 20 mM Tris-HCl, pH 7.6, 75 mM KCl, 3 mM CaCl$_2$, 1 mM dithiothreitol (DTT), 0.1% Nonidet P-40 (NP40), 8% glycerol.
2. MNase stop buffer: 20 mM Tris-HCl, pH 7.6, 10 mM EGTA, 0.5% SDS
3. 10X TPE buffer: 48.44 g Tris base, 40 mM final 1X concentration, 41.4 g NaH$_2$PO$_4$.H$_2$O to 30 mM (1X), 3.72 g EDTA to 1 mM (1X).
   Dissolve in water and make the final volume to 1 L. The final pH should be about 8.2 without adjustment.
4. DNase I buffer: 20 mM Tris-HCl, pH 7.6, 50 mM KCl, 5 mM MgCl$_2$, 1 mM DTT, 0.1% NP40, 8% glycerol.
5. DNase I stop buffer: 20 mM Tris-HCl, pH 7.6, 10 mM EDTA, 0.5% SDS.

3. Methods

3.1. Preparation of Xenopus Oocytes

1. Cut the isolated ovary into small pieces with scissors.
2. Wash oocytes in a 50-mL Falcon tube extensively with 1X OR2 buffer 5 to 6 times.
3. Weigh out 20 mg of collagenase (type I, Sigma) in a 50-mL conical tube and add 10 mL of 1X OR2 buffer to dissolve. Add oocytes to this tube with collagenase. This yields approximately a 2 mg/mL final concentration of collagenase.
4. Place the Falcon tube on shaker platform and shake for 1 to 2 h. Check after 1 h. Oocytes should completely separate from one another. However, do not treat with collagenase too long. This will damage oocytes.

5. Wash oocytes 6 times with 1X OR1 buffer.

6. Wash oocytes twice with 1X MBSH plus antibiotics (50 µg/mL penicillin and 50 µg/mL streptomycin).

7. Transfer oocytes to a petri dish containing 1X MBSH plus antibiotics and leave at 18°C for at least 3 h or overnight for recovery from collagenase treatment. Select healthy stage VI oocytes (size about 1200 µm) (*see* **Note 5**).

3.2. Microinjection of Xenopus Oocytes

Injection of reporter DNA into nucleus (germinal vesicle [GV]) of *Xenopus* oocytes is technically demanding and requires a lot of practice, more so than injection of mRNAs encoding TR and RXR into cytoplasm of *Xenopus* oocytes. Since the nucleus has a volume of approximately 50 nL, we usually inject DNA, no more than 18.4 nL, into each nuclei to reduce leakage of DNA into cytoplasm. Since the nucleus of *Xenopus* oocytes are located closer to the animal pole (pigmented side), the oocytes first need to be oriented with the animal hemisphere (with brown or dark pigment) facing up in the Nitex mesh or in an injection plate by using forceps under microscope. The injection needle is pushed into oocytes from the center point of the animal pole in parallel with the axis of animal-vegetable pole, and penetrate only about 1/5 to 1/4 distance of the oocyte to ensure a successful injection of DNA into the nucleus (**Fig. 1A**). To make single-stranded plasmid DNA (ssDNA) for injection, the promoters interested have to be cloned into M13 or phagemid based vectors such as pBluescript II.

3.3. Analysis of Transcription from Reporter DNA

3.3.1. RNA Purification

1. Select 15–20 injected healthy oocytes from each group after overnight incubation. Wash the oocytes with MBSH buffer once. Homogenize the oocytes in 200 µL of oocyte homogenization buffer by pippeting up and down several times with a p200 pipeter until no large chunk of oocyte debris is left (*see* **Note 6**).

2. While transcription from the injected DNA is usually the primary concern of the experiment, it is also desirable to recover DNA from injected oocytes to analyze chromatin remodeling by supercoiling assay or to check variation of DNA injection by Southern or slot blot. For this purpose, half of the sample (100 µL) from each homogenate is transferred to a 1.5-mL tube with 200 µL of DNA recovery buffer and used for recovery of DNA (*see* **Subheading 3.3.3.** DNA Purification), (*see* **Note 7**).

3. To purify RNA from each group of injected oocytes, add 400 µL of RNAzol (Tel-Test) and 40 µL of chloroform to 100 µL of the remaining lysate. Vortex mix vigrously and leave on ice for 15 min. Although we routinely use RNAzol for this purpose, other reagents such as TRIzol (Gibco-BRL) work equally well.

4. Centrifuge for 10 min at top speed (13,000 rpm, 16,000*g*) with a microfuge. Transfer 300 µL of clean supernatant to a new tube. Extract the supernatant once with an equal volume of chloroform and then precipitate RNA with addition of 0.7 vol of isopropanol. Store at –20°C for 20 min.

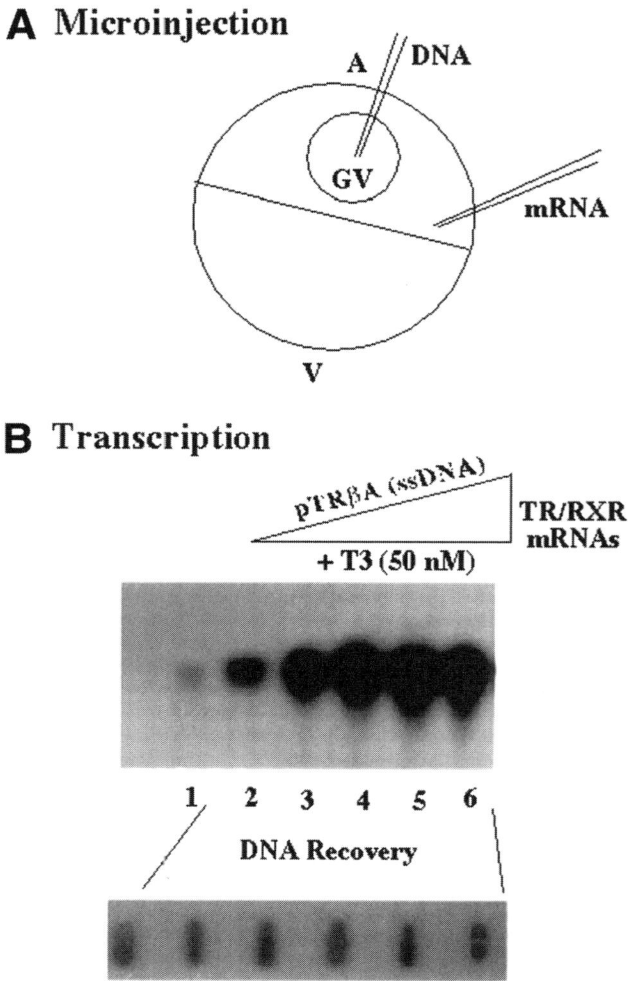

Fig. 1. (**A**) Diagram illustrates the microinjection of DNA and RNA. A, animal hemisphere; V, Vegetable hemisphere; GV, germinal vesicle (nucleus). (**B**) Liganded TR/RXR activates transcription from the repressive chromatin template assembled via the replication-coupled chromatin assembly pathway. Groups of 20 oocytes were first injected into cytoplasm with an increasing amount of TR/RXR mRNAs (1.2, 3.7, 11.1, 33.3, and 100 ng/μL, lanes 2–6) and then injected with ssDNA of pTRβA (100 ng/ μL, 18.4 nL/oocyte) into nucleus. The oocytes were incubated at 18°C overnight with T3. The transcription was analyzed by primer extension using a ^{32}P-labeled TRβA specific primer. To make sure that injection of ssDNA is uniform for each group of samples, DNAs were recovered from each group and analyzed by slot-blot hybridization with a probe from the TRβA promoter containing the TRE site.

5. Centrifuge in a microfuge at top speed (16,000g) for 10 min. Discard supernatant and rinse the pellet with 160 μL of cold 75% ethanol. Air-dry the pellets at room temperature.
6. Resuspend the pellet with diethyl pyrocarbonate (DEPC)-treated water in a volume of 3 to 4 μL for each oocyte. The RNA is now suitable for primer extension analysis.

3.3.2. Primer Extension Analysis

While levels of transcription from injected reporter can be analyzed by Northern blot, RNase protection assay, or primer extension analysis, we routinely use primer extension, since it is more convenient and less time-consuming. To minimize the possibility of premature termination during primer extension, specific oligonucleotide primer with a length of 25–30 nucleotides (nt) is designed to be located about 100–150 nt downstream of the transcription start site(s).

1. To anneal one or more ^{32}p-labeled primer(s) to specific RNA products, add and mix following components in an eppendorf tube (*see* **Note 8**): 2 μL 5X annealing buffer, 1 μL one or more labeled primer (10,000 counts per minute [cpm]) (**Note 5**), 4 μL RNA (equvailent to RNA from one oocyte), 0.1 μL RNase inhibitor, and H$_2$O to a total of 10 μL.
2. Incubate the tube at 65°C for 10 min, 55°C for 10 min, and 45°C for 25 min.
3. Add to each reaction: 4 μL 10X RT buffer, 1 μL 25 mM dNTP mixture, 1 μL 0.1 M DTT, 1 μL 1 μg/uL actinomycin D, 1/8 μL Superscript II, and 23 μL H$_2$O. Mix and incubate at 42°C for 1 h.
4. Precipitate primer extension products by addition of 4 μL of 3 M NaOAC, pH 5.2, and 100 μL of ethanol. Mix and leave in –20°C for 30 min.
5. Centrifuge at top speed in a microfuge for 10 min to pellet DNA. Discard supernatant. Rinse pellets with 75% ethanol once and air-dry.
6. Resuspend the pellet with 2 uL of H$_2$O and 3 μL of formamide loading buffer. Vortex to dissolve pellets. Heat at 90°C for 2 min to denature primer extension products before loading to a sequencing gel (6%).
7. After gel running, transfer the gel to a piece of 3MM filter paper. Cover the gel with saran wrap. Do autoradiograph with or without drying the gel (**Note 9**).

3.3.3. DNA Purification

1. To the mixture from **Subheading 3.3.1.**, **item 2.**, add 1 μL of RNase A (10 mg/mL) and incubate at 37°C for 1 h. (**Note 10**).
2. Add 3 μL of proteinase K (10 mg/mL) and incubate at 55°C for at least 2 h or overnight. Based on our experience, treatment with proteinase K is absolutely necessary for successful recovery of DNA, possibly due to the presence of large amount of storage proteins in *Xenopus* oocytes.
3. Add 30 μL of 3 M NaOAC, pH 5.2. Extract with 350 μL of phenol–chloroform once. Centrifuge at top speed for 5 min. Carefully transfer supernatant to a new tube and extract with an equal volume of chloroform once. Precipitate DNA by addition of a 0.7 vol of isopropanol and leave at –20 freezer for at least 30 min.

4. Centrifuge at 13,000 rpm (16,000*g*) for 15 min to collect DNA. After centrifugation, discard the supernatant and wash the pellet with 160 µL of cold 75% ethanol. Resuspend DNA in distilled water in a volume of 5 µL per oocytes. The DNA is now suitable for the supercoiling assay, as well as for Southern or slot hybridization to check the recovery of DNA.

3.4. Analysis of Chromatin Structure of Reporter DNA in Xenopus Oocytes

3.4.1. MNase Assay of Chromatin Structure Assembled in Xenopus Oocytes

1. Inject groups of oocytes (15–20) with mRNAs encoding TR/RXR and with ssDNA or dsDNA of the interested plasmid (pTRβA). As experienced, dsDNA is injected as a concentration of 100 ng/µL and ssDNA as 50 ng/ µL in a volume no more than 18.4 nL/oocyte. Incubate oocytes at 18°C overnight.
2. Collect the healthy oocytes in each group and rinse the oocytes with 500 µL of MBSH buffer once.
3. Add 200 µL of MNase buffer and homogenize the oocytes by pippeting. Label four tubes (A, B, C, and D) for each group of oocytes. Add 60 µL of the lysates to tube A, and 40 µL to tubes B, C, and D, respectively.
4. Add 2.5 µL of 5 U/µL MNase to tube A and mix well. Transfer 20 µL of the mixture out of tube A to tube B and mix well. In turn, transfer 20 µL of the sample from tube B to tube C and mix them well. Repeat it from tube C to tube D. This will give a series of MNase digestion with a 3-fold dilution of MNase from one tube to next. Incubate at room temperature for 20 min.
5. Stop MNase digestion by addition of 200 µL MNase stop.
6. Add 1 µL of RNase A (10 mg/mL). Incubate at 37°C for 1 h.
7. Add 3 µL proteinase K to each tube and incubate at 55°C for 2 to 3 h or overnight.
8. Add 30 µL of 3 *M* NaOAC, pH 5.2. Extract with 300 µL of phenol–chloroform once. Centrifuge at top speed (16,000*g*) for 5 min. Carefully transfer supernatant to a new tube and extract with an equal vol of chloroform once. Precipitate DNA by addition of a 0.7 vol of isopropanol and leave at –20°C freezer for at least 30 min.
9. Centrifuge at 13,000 rpm (16,000*g*) for 15 min to collect DNA. After centrifugation, discard supernatant and wash the pellet with 160 µL of cold 75% ethanol. DNA was then resuspended in 15 µL of distilled water.
10. Prepare a 1.5% agarose gel in 1X TBE buffer. Add 3 µL of 6X gel loading buffer to DNA and load to the agarose gel. Run at 80–100 V for 2 to 3 h.
11. Blot DNA to a nylon membrane and perform Southern hybridization according to the basic protocol in **ref. *20*** (*see* **Note 11**).

3.4.2. Supercoiling Assay

The disruption of chromatin structure can be also detected by supercoiling assay. This assay is based on the observation that in vitro the assembly of each nucleosome into a circular plasmid DNA will introduce one negatively super-helical turn *(21)*. If a transcription factor has the capacity to exert disruption of chromatin, it will result in the change of the number of superhelical turns. The plasmid molecules with different superhelical turns can be resolved in agarose gel containing chloroquine.

1. Make a 1–1.2% agarose gel in 1X TPE buffer containing 90 µg/mL of chloro-quine diphosphate (Sigma; cat no. C-6628). Add chloroquine when agarose solution is not very hot (approx 70°C).
2. Prepare 1 L of 1X TPE buffer with 90 µg/mL of chloroquine.
3. Add agarose gel loading buffer without bromophenol blue to DNA samples recovered from injected oocytes (*see* **Subheading 3.3.3.** DNA Preparation) (bro-mophenol blue forms precipitates with chloroquine). As a marker, xylene cyanole FF (0.01%) can be added to the gel loading buffer. As experienced, DNA recov-ered from 2 to 3 injected oocytes is sufficient for supercoiling assay detected by Southern hybridization.
4. Load DNA samples to gel. Circulate TPE electrophoresis buffer using a Pharmacia P-1 pump with a high flow rate. It is necessary to circulate the buffer because the TPE buffer has a low capacity. Cover the gel apparatus with alumi-num foil or run the gel in the dark (chloroquine is light-sensitive). Run the gel at 35 V overnight (it depends on the size of the plasmid analyzed. For a 4.5-kb TRβA construct, it take about 12–14 h).
5. After electrophoresis, wash the gel twice in 500 mL water for 1 to 2 h to remove chloroquine.
6. Treat the gel with 500 mL of 0.25 *M* HCl for 30 min.
7. Rinse the gel with water and then treat with 0.5 *M* NaOH, 1.5 *M* NaCl for 30 min, followed by treatment with 0.5 *M* Tris-HCl, pH7.5, 1.5 *M* NaCl for 30 min.
8. Transfer DNA to a piece of nylan membrane with 20X standard saline citrate (SSC) according to "Current Protocol in Molecular Biology" *(20)*.
9. Supercoiling pattern can be detected with any probes derived from the constructs to be analyzed (**Note 12**).

3.4.3. DNase I Hypersensitive Site Assay of Chromatin Assembled in Xenopus Oocytes

DNase I hypersensitive site assay is perhaps the most widely used assay for detection of changes in chromatin structure. Studies of many spatially or temporally regulated genes, such as β-globin, have revealed the correlation of formation of DNase I hypersensitive sites with the status of transcription *(22)*. This assay has also been used extensively in analyses of chromatin structure in injected *Xenopus* oocytes *(24)*. Although the mechanism for the formation of DNase I hypersensitive sites is still not fully understood, the detection of DNase I hypersensitive sites are widely interpreted as results of chromatin remodeling induced by the binding of transcription factors.

1. Inject 15–20 oocytes with DNA and with or without mRNA encoding TR/RXR (100 ng/ µL, 27.6 nL/oocyte) and treated with or without 50 n*M* T3. Incubate the oocytes overnight at 18°C incubator.
2. Collect healthy oocytes and wash the oocytes once with 500 µL of MBSH buffer.
3. To 15–20 oocytes, add 240 µL of DNase I buffer. Homogenize the oocytes by pipeting.
4. Distribute 60 µL of the lysates to four tubes labeled as A, B, C, and D, respectively. Dilute DNase I (Gibco-BRL, 133.9 U/ µL) 1/3 and1/9 in TE buffer right before use. Add 0.5 µL of undiluted DNase I to tube A, 1/3 diluted DNase I to B, and 1/9 diluted DNase I to C. No DNase I is added to tube D which will serve as a no-digestion control. Incubate at room temperature for 10 min.
5. Stop DNase I digestion by addition of 200 µL of DNase I stop buffer. Add 1 µL of RNase A (10 mg/mL). Incubate at 37°C for 1 h.
6. Add 3 µL of proteinase K (10 mg/mL) to each tube. Incubate at 55°C for 2 to 3 h.
7. Add 30 µL of NaOAC, pH 5.4 to each tube. Extract with 300 µL of phenol–chloroform once. Carefully transfer the supernatant to a new tube and extract with chloroform once more. Precipitate DNA by addition of 0.7 V of isopropanol. Wash with 70% ethanol once.
8. Resuspend DNA in 43 µL of water. Add 2 µL of the appropriate restriction enzyme (for indirect end-labeling) and 5 µL of 10X restriction buffer according to the restriction enzyme used for digestion. Incubate at 37°C for at least 3 h to overnight.
9. Resolve DNA samples with a 1–1.2% agarose gel in 1X TBE. Transfer DNA to a Nylon membrane and carry out the standard Southern blotting using the probe one end at the restriction site and the other end within the suspected hypersensitive sites (**Note 13**).

3.5. Concluding Comments

It has become clear over last several years that *Xenopus* oocyte serves as an excellent model system for study of transcriptional regulation in the context of chromatin *(15,24,25)*. By exploiting this system, we have been able to demonstrate that chromatin assembly is important for both efficient repression and activation by unliganded TR/RXR and liganded TR/RXR, respectively, and that liganded TR/RXR, can induce targeted disruption of chromatin *(15,26)*. In addition, we have shown recently that the repression by unliganded TR/RXR requires the involvement of histone deacetylase activity, since treatment of oocytes with trichostatin A (TSA), a specific inhibitor of histone deacetylases, blocks repression by unliganded TR/RXR *(27)*. Over the last several years, it has also become clear that both repression by unliganded TR/RXR and activation by liganded TR/RXR require cofactors termed corepressors and coactivators, respectively. More and more corepressors and coactivators have been identified, and the mechanisms by which they modulate nuclear hormone receptor function have begun to emerge. Clearly, many prominent coactivators, such as CBP/p300, SRC1/ACTR, and PCAF/GCN5 contain histone acetyltransferase activity, providing a strong connection between acetylation of chromatin with transcriptional activation *(7,28)*. On the other hand, corepressors SMRT and NCoR are found to associate with multiple histone deacetylases, providing a strong connection between histone deacetylation and transcriptional repression *(7,29,30)*. The challenge in future is to dissect the roles of multiple coactivators and corepressors in the repression or activation process. In this regard, the development of the chromatin immunoprecipitation (ChIP) assay *(31)* will allow us to gain insight into the detailed modification of histones, as well as what corepressors or coactivators are actually involved in repression and activation by unliganded and liganded TR/RXR, respectively.

Another important feature regarding the chromatin assembled in *Xenopus* oocytes is the positioning of nucleosome. The precise positioning of nucleosome has been shown to play an important role in transcriptional regulation. Analyses of *Xenopus* TRβA promoter in oocytes revealed that the promoter is assembled into positioned nuclesomes *(17,32)*. In future, it is of interest to determine whether the TRβA promoter is also assembled into positioned nucleosomes in cells.

In essence, *Xenopus* oocyte is a powerful model system that offers unique opportunity to study transcriptional regulation by nuclear hormone receptors as well as other transcriptional factors in the context of chromatin.

4. Notes

1. One additional advantage offered by *Xenopus* oocytes is that one can inject DNA or mRNAs in different order for different experimental purpose. For example, if ssDNA reporter is injected first and the mRNA encoding a transcription factor is injected 2 to 3 h later, the injected DNA will be assembled into chromatin before the transcription factor is synthesized in oocytes. This provides an opportunity to test whether a transcription factor has the capacity to bind to and regulate transcription of its target gene preassembled into chromatin. Using this strategy, we have demonstrated that TR/RXR have the capacity to bind to its recognition sequence (TRE) preassembled into chromatin and repress or activate transcription depending on the absence or the presence of T3 *(15)*.

2. We prefer Nanoject II from Drummond Scientific Company since it allows injection of variable volumes ranging from 4.6–73.6 nL. For injection of mRNAs, a volume up to 45 nL can be injected, although we usually inject no more than 27.6 nL of mRNAs into a single oocyte. If injection of more mRNAs is desirable, it is recommended to prepare a higher concentration of mRNA rather than injection a larger volume of mRNAs.

3. Based on our experience, almost any genes cloned into vectors containing a viral promoter, such as SP6, T7, or T3, can be transcribed into mRNAs in vitro using an in vitro transcription kit (mMESSAGE mMACHINE™). Before in vitro transcription, the expression construct should be linearized by using an appropriate restriction enzyme, which cut downstream of 3' end of the gene or poly(A) tail.

These kits allow synthesis of approx 10–20 µg of mRNA containing a 5' cap from about 1 µg of template DNA. In addition to its role in mRNA stabilization, the 5' cap is known to increase the translation efficiency of injected mRNAs in *Xenopus* oocytes at least 5-fold. Although the effect of poly(A) tail on the stability of mRNAs and efficiency of translation is still not well defined, it is generally believed that addition of a poly(A) tail to mRNA will help stability and improve translation efficiency of the mRNA. For this reason, we routinely clone all genes for oocyte expression into pSP64poly(A) vector. Usually, the levels of protein expression in oocytes are correlated to the amount of the mRNA injected, until the translation capacity of the oocytes is saturated.

4. A good protocol for preparing ssDNA from M13 or phagemid constructs can be found in Current Protocol in Molecular Biology (1.15.1-5) *(33)*. The quality of double-stranded plasmid DNA prepared using a variety of miniprep or large-scale isolation kits usually are suitable for oocyte injection. The DNAs for injection can be dissolved in distilled water. Since the levels of transcription vary from one promoter to another, the optimal concentration of DNA injected should be determined empirically. We routinely inject ssDNA or dsDNA in a concentration ranging from 10–100 ng/µL in a volume of 18.4 nL/oocyte. A more detailed discussion on how to inject mRNA and DNA into *Xenopus* oocytes can be found in "Methods in Cell Biology" *(23)*.

5. Treatment with collagenase causes some damage to oocytes and may not be of the choice if injection of limited number of oocytes is needed. In this case, healthy stage VI oocytes can be dissected manually from the isolated ovary with fine forceps. However, if a large number of oocytes are to be injected and, in particular, the injection of DNA into the nucleus of oocytes is involved, we recommend treatment with collagenase. The collagenase treatment not only frees individual oocytes, but also serves to remove follicular membrane, which is hard to penetrate, from the oocytes, thus facilitating the subsequent microinjection.

6. Based on our experience, transcription from injected reporters can be detected 2 to 3 h after injection and peaks around 8–10 h. To minimize variation due to microinjection, each sample should contain about 15–20 injected oocytes. Based on our experience, variation due to injection is negligible once microinjection skill is improved.

7. To further minimize variation, one can also select a primer extension product from an endogenous mRNA as internal loading control. We routinely use histone H4 mRNA as an internal loading control *(16)*.

8. The primer is end-labeled by using [r-^{32}p]ATP and T4 polynucleotide kinase and purified by Sephadex G-50 spin column according to manufacturer's instruction (Promega).

9. A representative result of transcription activation induced by liganded TR/RXR is shown in **Fig. 1B**. In this case, TRβA promoter was assembled into repressive chromatin via injection of ssDNA. Expression of TR/RXR through injection of an increasing amount of TR and RXR mRNA mixture led to strong T3-dependent activation in a dose-dependent manner.

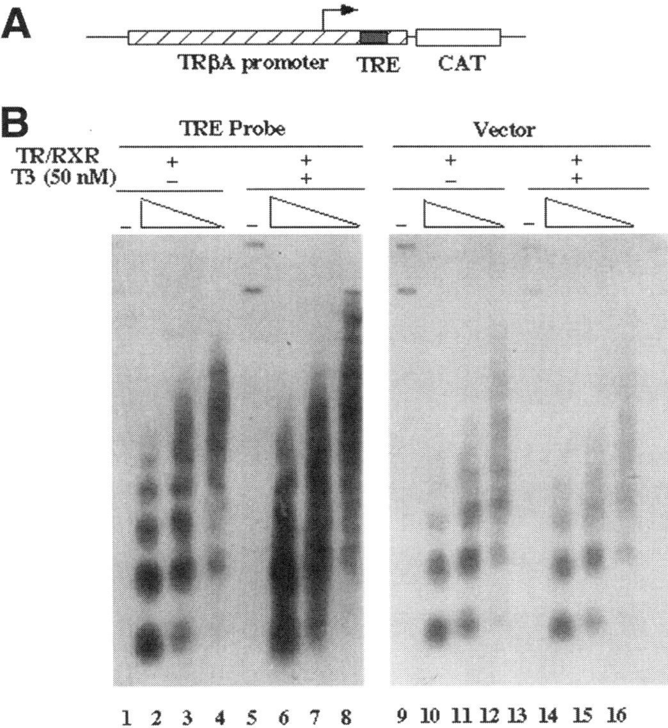

Fig. 2. Liganded TR/RXR instigates extensive chromatin disruption. (A) The diagram shows the structure of the pTRbA construct and the location of the TRE probe used for hybridization. The TRE probe is a 96-bp fragment (from +218 to +314) generated by polymerase chain reaction (PCR), which covered the TRE site. The vector probe is ^{32}P-labeled pBluescript II SK(+). The oocytes were injected with ssDNA (50 ng/μL, 18.4 nL/oocyte) and TR/RXR mRNAs (100 ng/μL each, 27 nL/oocyte) and treated with or without hormone and processed for MNase assay. The amounts of MNase used are zero (lanes 1 and 5), 12.5 U (lanes 2 and 6), 4.2 U (lanes 3 and 7), and 1.3 U (lanes 4 and 8). The same filter was hybridized sequentially with random-primed labeled vector probes as indicated.

10. One can also recover the injected DNA from oocytes and use them to check for equal injection of DNA into different groups of oocytes or for chromatin remodeling by supercoiling assay. As an example, the DNA recovered from each group of the injected oocytes (*see* **Fig. 1B**) was examined by slot blot using TRE probe (*see* **Fig. 2A**), and the result was shown at the lower panel of **Fig. 1B**.

11. The result of chromatin disruption induced by TR/RXR as revealed by MNase assay is shown in **Fig. 2B**. Clearly, the chromatin disruption as revealed by loss of clear nucleosomal array and appearance of sub-nucleosomal fragments is observed

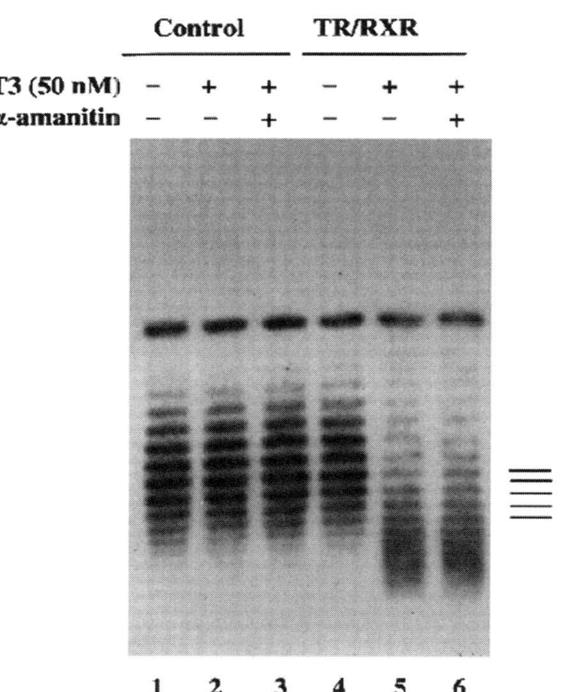

Fig. 3. The DNA topology assay indicates that liganded TR/RXR also induces extensive chromatin disruption and that this chromatin disruption is not the by-product of ongoing transcription. Groups of 20 oocytes were injected with ssDNA and TR/RXR mRNAs as in **Fig. 2** and treated with or without hormone as indicated. To examine whether the disruption of chromatin structure is dependent on the process of transcription, a-amanitin was co-injected with ssDNA at a concentration of 10 mg/mL. Under this condition, the T3-dependent transcription activation was completely inhibited (data not shown). The DNA was purified from each group, and the topological status of the DNA was analyzed using chloroquine agarose gel as described for the supercoiling assay (**Subheading 3.4.2.**). The mean difference in superhelical density between lanes 4 and 5 is approx 5–6 helical turns. The top band in each lane represents the nicked form of plasmid.

only in the presence of T3, indicating that liganded TR/RXR has the ability to actively recruit chromatin remodeling activities. In addition, as revealed by sequential hybridization using labeled vector, the chromatin disruption was not observed in the vector region, indicating that the chromatin disruption was centered around the TRE site *(16)*.

12. A representative result of supercoiling assay for chromatin remodeling induced by liganded TR/RXR is shown in **Fig. 3**. Consistent with the result of MNase assay shown in **Fig. 2**, TR/RXR induces chromatin disruption only in the presence of T3. Thus, the result of supercoiling assay is consistent with that of MNase

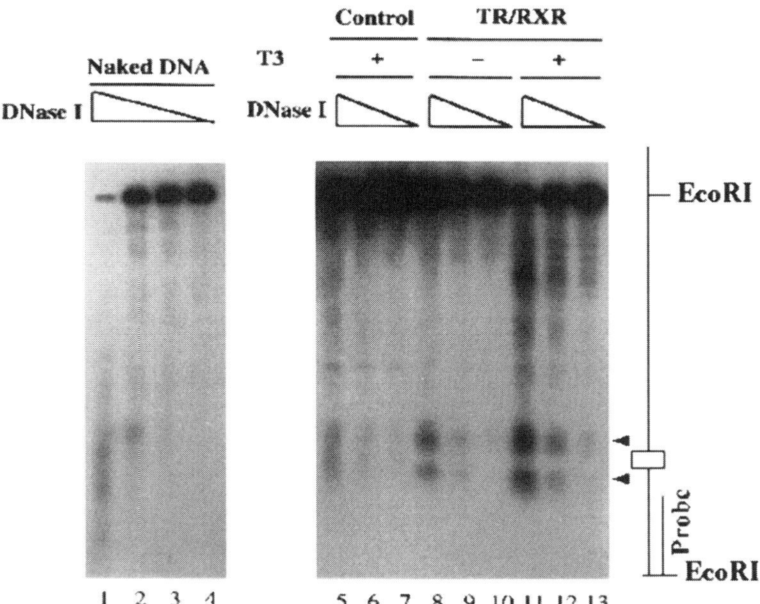

Fig. 4. DNase I hypersensitive assay revealed distinct chromatin structure induced by unliganded and liganded TR/RXR. The groups of oocytes were injected with or without TR/RXR mRNAs and ssDNA of pTRbA, as described in **Fig. 2**, and treated with or without T3 as indicated. After overnight incubation, the groups of oocytes were collected and used for the DNase I sensitivity assay. The DNase I concentrations used in digestion were 40 U (lanes 5, 8, and 11), 20 U (lanes 6, 9, and 12) and 10 U (lanes 7, 10, and 13). The naked plasmid controls were treated with less of DNase I (0.4, 0.2, 0.1, and 0.05 U for lanes 1–4, respectively). Also indicated are the positions of EcoRI sites and the TRE. The probe used here for hybridization is the CAT sequence indicated in **Fig. 2**. The asterisk indicates the position of the DNase I hypersensitive sites detected in the absence of T3. Note that liganded TR/RXR induced stronger cleavage on both sites detectable in the absence of T3 and additional DNase I hypersensitive sites in the TRbA promoter.

assay. Importantly, the supercoiling assay has the advantage to quantitatively reveal the degree of chromatin disruption. It appears that approx 5 to 6 nucleosomes are disrupted upon treatment with T3. Given the fact that the TRβA promoter contains 2 to 3 separate TREs (*16*), we estimate that each TRE can roughly confer the disruption of two nucleosomes.

13. A representative result of DNase I hypersensitive site assay from our study of TR/RXR is shown in **Fig. 4**. It is interesting to compare the results of the DNase I hypersensitive assay and MNase assay. As shown in **Fig. 2**, no detectable change in chromatin structure is observed in the presence of TR/RXR but the absence of

T3. The extensive disruption of chromatin structure is only observed in the presence of both TR/RXR and T3. In contrast, a clear DNase I hypersensitive site can be detected in the presence of TR/RXR but the absence of T3. Addition of T3 led to the increased sensitivity and the formation of additional hypersensitive sites. Thus, the comparison of the results from these two assays suggests that the DNase I hypersensitive sites observed in the absence of T3 reflect the binding of TR/RXR to chromatin. Indeed, the binding of unliganded TR/RXR to TRE in chromatin was confirmed by DNase I footprinting analyses on the TRβA promoter assembled into chromatin in *Xenopus* oocytes or into nucleosome in vitro. Such binding in the absence of T3 induced minimal, if any, change of chromatin structure, since no change in chromatin structure was detected by MNase assay. On the other hand, the extensive chromatin disruption that occurred in the presence of T3, can be detected by all three assays listed here. Thus, it is clear that each of the three assays has its advantage and limitation, and a comprehensive view of how a transcription factor binds to and induces the change of chromatin structure can be better understood if all three assays are used.

References

1. Felsenfeld, G., Boyes, J., Chung, J., Clark, D., and Studitsky, V. (1996) Chromatin structure and gene expression. *Proc. Natl. Acad. Sci. USA* **93,** 9384–9388.
2. Wolffe, A. P. (1997) Histones, nucleosomes and the roles of chromatin structure in transcriptional control. *Biochem. Soc. Trans.* **25,** 354–358.
3. Grunstein, M. (1997) Histone acetylation in chromatin structure and transcription. *Nature* **389,** 349–352.
4. Kingston, R. E. and Narlikar, G. J. (1999). ATP-dependent remodeling and acetylation as regulators of chromatin fluidity. *Genes Dev.* **13,** 2339–2352.
5. Peterson, C. L. (1996) Multiple SWItches to turn on chromatin? *Curr. Opin. Genet. Dev.* **6,** 171–175.
6. Tsukiyama, T. and Wu, C. (1997) Chromatin remodeling and transcription. *Curr. Opin. Genet. Dev.* **7,** 182–191.
7. Glass, C. and Rosenfeld, M. G. (2000) The coregulator exchange in transcriptional functions of nuclear receptors. *Genes Dev.* **14,** 121–141.
8. Wolffe, A. P., Wong, J., and Pruss, D. (1997) Activators and repressors: making use of chromatin to regulate transcription. *Genes Cells* **2,** 291–302.
9. Wolffe, A. P. (1997). Chromatin remodeling regulated by steroid and nuclear receptors. *Cell Res.* **7,** 127–142.
10. Baniahmad, A., Steiner, C., Kohne, A. C., and Renkawitz, R. (1990) Modular structure of a chicken lysozyme silencer: involvement of an unusual thyroid hormone receptor binding site. *Cell* **61,** 505–514.
11. Glass, C. K., Lipkin, S. M., Devary, O. V., and Rosenfeld, M. G. (1989) Positive and negative regulation of gene transcription by a retinoic acid-thyroid hormone receptor heterodimer. *Cell* **59,** 697–708.
12. Laskey, A., Mill, D., and Morris, N. (1977) Assembly of SV40 chromatin in a cell-free system from *Xenopus* eggs. *Cell* **10,** 237–243.

13. Almouzni, G. and Wolffe, A. P. (1993) Replication-coupled chromatin assembly is required for the repression of basal transcription in vivo. *Genes Dev.* **7,** 2033–2047.

14. Wormington, M. (1991) Preparation of synthetic mRNAs and analyzes of translational efficiency in microinjected *Xenopus* oocytes. *Methods Cell Biol.* **36,** 167–183.

15. Wong, J., Shi, Y. B., and Wolffe, A. P. (1995) A role for nucleosome assembly in both silencing and activation of the *Xenopus* TR beta A gene by the thyroid hormone receptor. *Genes Dev.* **9,** 2696–2711.

16. Wong, J., Shi, Y. B., and Wolffe, A. P. (1997a) Determinants of chromatin disruption and transcriptional regulation instigated by the thyroid hormone receptor: hormone-regulated chromatin disruption is not sufficient for transcriptional activation. *EMBO J.* **16,** 3158–3171.

17. Wong, J. and Shi, Y. B. (1995) Coordinated regulation of and transcriptional activation by *Xenopus* thyroid hormone and retinoid X receptors. *J. Biol. Chem.* **270,** 18,479–18,483.

18. Wu, M. and Gerhart, J. (1991) Raising Xenopus in the laboratory. *Methods Cell Biol.* **36,** 3–18.

19. Smith, L. D., Xu, W., and Varnold, R. L. (1991) Oogenesis and oocyte isolation. *Methods Cell Biol.* **36,** 45–60.

20. Brown, T. (2000) Analysis of DNA sequences by blotting and hybridization, in *Current Protocols in Molecular Biology* (Ausobel, F. M., Breut, R., Kingston, K. E., et al., eds.) pp. 2.9.1–2.9.6.

21. Clark, D. J. and Wolffe, A. P. (1991) Superhelical stress and nucleosome-mediated repression of 5S RNA gene transcription in vitro. *EMBO J.* **10,** 3419–3428.

22. Felsenfeld, G. (1993) Chromatin structure and the expression of globin-encoding genes. *Gene* **135,** 119–124.

23. Kay, B. K. (1991) Injection of oocytes and embryo. *Methods Cell Biology* **36,** 663–669.

24. Landsberger, N. and Wolffe, A. P. (1997) Remodeling of regulatory nucleoprotein complexes on the *Xenopus* hp70 promoter during meiotic maturation of the *Xenopus* oocyte. *EMBO J.* **16,** 4361–4373.

25. Belikov, S., Gelius, B., Almouzni, G., and Wrange, O. (2000) Hormone activation induces nucleosome positioning in vivo. *EMBO J.* **19,** 1023–1033.

26. Wong, J., Li, Q., Levi, B. Z., Shi, Y. B., and Wolffe, A. P. (1997b) Structural and functional features of a specific nucleosome containing a recognition element for the thyroid hormone receptor. *EMBO J.* **16,** 7130–7145.

27. Wong, J., Patterton, D., Imhof, A., Guschin, D., Shi, Y. B., and Wolffe, A. P. (1998) Distinct requirements for chromatin assembly in transcriptional repression by thyroid hormone receptor and histone deacetylase. *EMBO J.* **17,** 520–534.

28. Wolffe, A. P., and Pruss, D. (1996) Targeting chromatin disruption: transcription regulators that acetylate histones. *Cell* **84,** 817–819.

29. Wolffe, A. P. (1996) Histone deacetylase: a regulator of transcription [comment]. *Science* **272,** 371–372.

30. McKenna, N. J., Xu, J., Nawaz, Z., Tsai, S. Y., Tsai, M. J., and O'Malley, B. W. (1999) Nuclear receptor coactivators: multiple enzymes, multiple complexes, multiple functions. *J. Steroid Biochem. Mol. Biol.* **69,** 3–12.

31. Solomon, M. J., Larsen, P. L., and Varshavsky, A. (1988) Mapping DNA-protein interactions in vivo with formaldehyde: Evidence that histone H4 is retained on a highly transcribed gene. *Cell* **53,** 937–947.
32. Guschin, D., Chandler, S., and Wolffe, A. (1998) Asymmetric linker histone association directs the asymmetric rearrangement of core histone interactions in a positioned nucleosome containing a thyroid hormone response element. *Biochemistry* **37,** 8629–8636.
33. Greerstein, D., and Begmond, C. Preparing and Using M13-Derived Vectors, in: Current Protocols in Molecular Biology, (Ausobel, F. M., Brent, R., Kingston, K. E., et al. eds.), pp 1.15.1–1.15.8.

11

Gene Activation by Thyroid Hormone Receptor In Vitro and Purification of the TRAP Coactivator Complex

Joseph D. Fondell

1. Introduction

Thyroid hormone receptors (TRs) are capable of both activating and repressing transcription from genes bearing TR-binding elements (TREs). In general, TRs function as activators in the presence of thyroid hormone (T3) and repressors in the absence of T3 *(1)*. The ability of TRs to regulate transcription has been linked to their ability to recruit distinct types of transcriptional coregulatory factors, termed coactivators and corepressors, to target gene promoters *(2,3)*. Remarkably, yeast 2-hybrid and conventional expression library screens have identified numerous TR-binding coregulatory factors (reviewed in **refs.** *2* and *3*). The best characterized TR coactivators include members the p160/SRC family *(2–4)*. While the precise mechanism of action of the p160/SRC proteins is still being defined, their ability to associate with histone acetyltransferases (HATs) such as the CREB-binding protein (CBP) and p300, and the presence of intrinsic HAT activity in some family members indicates a functional role in chromatin modification *(2–4)*.

Another type of mutimeric coactivator complex was recently purified from T3-cultured mammalian cells in association with ligand-bound TR *(5)*. The multisubunit complex is comprised of at least 14 different polypeptides (termed TRAPs for TR-associated proteins) ranging in size from approx 15–240 kDa. In vitro transcription assays show that the TRAP complex significantly enhances TR-mediated transcription on nonchromatin DNA templates *(5,6)* and in the absence of TATA-binding protein (TBP)-associated factors (TAFs) *(6)*. In contrast to SRC/p160-CBP/p300 complexes, the TRAP protein com-

From: *Methods in Molecular Biology, Vol. 202: Thyroid Hormone Receptors: Methods and Protocols*
Edited by: A. Baniahmad © Humana Press Inc., Totowa, NJ

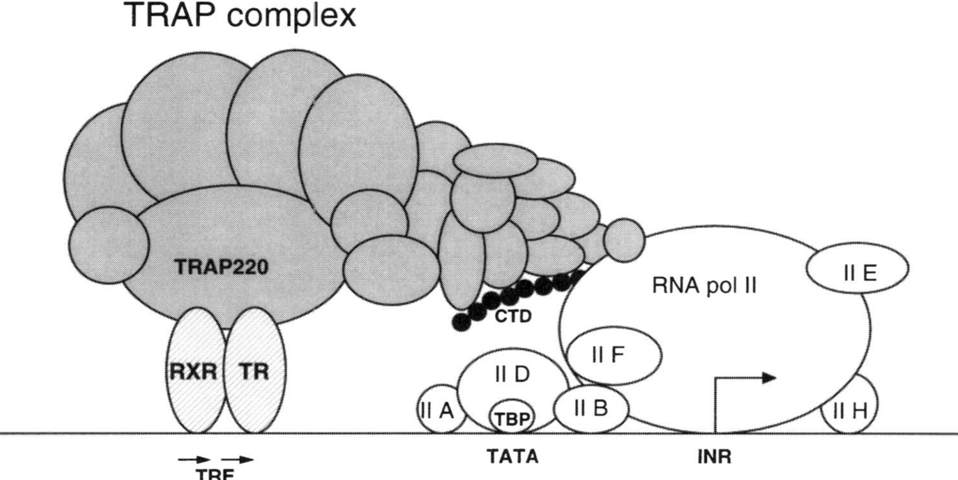

Fig. 1. A model for TR-dependent transcriptional activation mediated by the TRAP coactivator complex. Subunits of the multiprotein TRAP complex are represented as shaded ovals. TR-RXR heterodimers interact with the TRAP220 subunit in a T3-dependent manner *(10)* and target the TRAP complex to TREs. Recruitment of the TRAP complex to specific promoters serves to enhance the assembly or the activity of RNA polymerase II (RNA pol II) and the general transcription factors (TFII-A, -B, -D, -E, -F, and -H) at the core promoter elements (the TATA box [TATA] and initiator element [INR]). Functional interactions between specific TRAP subunits and the basal transcription apparatus may occur, at least in part, via the carboxy-terminal domain (CTD) (shown as small black spheres) of the largest subunit of RNA pol II *(11)*.

plex appears to lack HAT activity. These findings suggest that TRAPs facilitate a novel TR-coactivator pathway or step, distinct from those mediated by SRC/p160-CBP/p300 complexes, and likely involving a more direct influence on the basal transcription machinery (**Fig. 1**). Consistent with the view that TRAPs serve as a direct interface between TR and the basal apparatus, several TRAP subunits appear to be human homologs of yeast proteins found within Mediator, a large coactivator holocomplex directly associated with both transcriptional activators and the yeast RNA polymerase II holoenzyme *(7,8)*.

Evidence for a TRAP coactivator role in other nuclear receptor (NR) signaling pathways comes from the recent purification of a similar, if not identical, complex of cofactors (termed DRIPs), which associate with the vitamin D receptor (VDR) and stimulate VDR-mediated transcription in vitro *(9)*. A single TRAP subunit (TRAP220) has been shown to contact TR and VDR (as well numerous other NRs) in a ligand-dependent manner and is believed to anchor the TRAP complex to DNA-bound NRs (reviewed in **ref. 10**) (*see* **Fig. 1**). In addition, most (if not all) TRAP subunits have been identified in other metazoan transcriptional holocomplexes including NAT, SMCC, ARC, and CRSP (reviewed in **ref.** *11*). The recruitment of these various TRAP/ mediator-like complexes to specific gene promoters appears to be facilitated by a broad range of transcriptional activators unrelated to NRs, including p53, Sp1, and SREBP, and presumably involve target interactions with alternative TRAP subunits *(11)*.

The use of cell-free transcription assays provides a powerful method for dissecting the precise molecular steps of transcriptional regulation by TR. In vivo situations in which TR (in concert with coactivator complexes) functionally interacts with the basal transcription apparatus can be mimicked in vitro by incubating purified TR-coactivator complexes in either: (*i*) a reconstituted system composed of highly purified general transcription factors and RNA polymerase II; or (*ii*) a soluble protein extract prepared from mammalian cell nuclei. Specific TR-dependent transactivation in vitro can be subsequently measured from a TRE-linked promoter template. Similarily, a highly defined in vitro transcription system can also be used to investigate the detailed molecular mechanisms of TR-mediated repression (gene silencing).

This chapter describes a method for purifying the TR-TRAP coactivator complex from T3-cultured mammalian cells and further outlines an in vitro transcription assay, which can be used to investigate TR-TRAP activity. A retroviral gene transfer technique is used to establish a HeLa-derived cell line, which stably expresses a recombinant FLAG epitope-tagged TR (f:TR). Nuclear extract is prepared from T3-induced cells, and the f:TR-TRAP complex is immunopurified using anti-FLAG monoclonal antibodies conjugated to agarose beads. The f:TR-TRAP complex is then eluted from the resin with synthetic FLAG peptide. The experimental rational for this protocol is illustrated in **Fig. 2**. As a source of purified retinoid X receptor (RXR; the heterodimerization partner of TR) and control TR protein required for the in vitro transcription assay, a protocol for the purification of baculovirus-expressed f:TR and f:RXR from infected Sf9 cells is also outlined.

2. Materials

2.1. Generation of a Stable Epitope-Tagged TR-Expressing Hela Cell Line

1. pBABEneo-FLAG-TR plasmid DNA (*see* **Note 1**).
2. ψCRIP retroviral packaging cell line *(14)*.
3. HeLa S3 cells (American Type Culture Collection, Rockville, MD).
4. Dulbecco's modified Eagles's medium (DMEM) (Life Technologies, Gaithersburg, MD, cat. no. 11995-065).
5. Joklik's minimal essential medium (Life Technologies, cat. no. 22300-024).

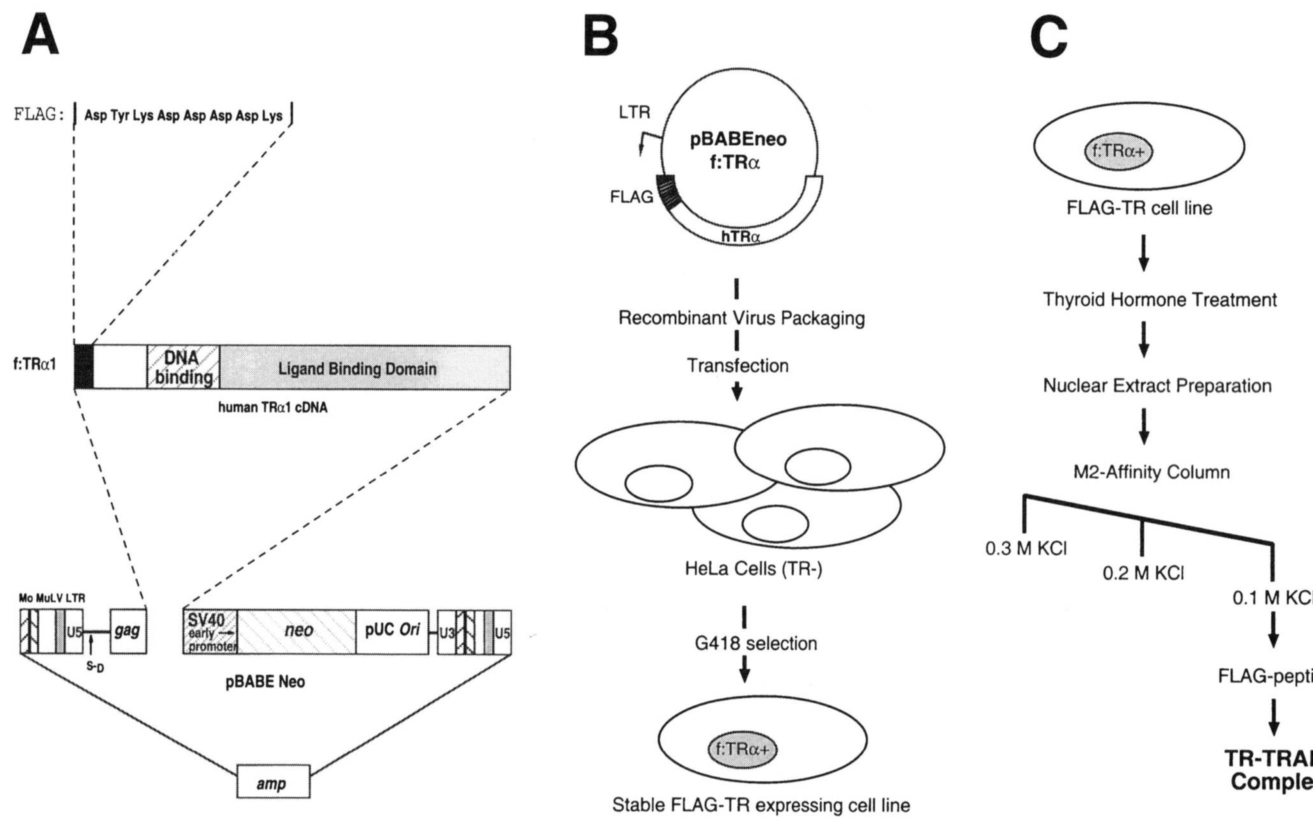

A
FLAG: Asp Tyr Lys Asp Asp Asp Asp Lys
f:TRα1
DNA binding
Ligand Binding Domain
human TRα1 cDNA
Mo MuLV LTR
U5
S-D
gag
SV40 early promoter
neo
pUC Ori
U3
U5
pBABE Neo
amp
B
LTR
pBABEneo f:TRα
FLAG
hTRα
Recombinant Virus Packaging
Transfection
HeLa Cells (TR-)
G418 selection
f:TRα+
Stable FLAG-TR expressing cell line
C
f:TRα+
FLAG-TR cell line
Thyroid Hormone Treatment
Nuclear Extract Preparation
M2-Affinity Column
0.3 M KCl
0.2 M KCl
0.1 M KCl
FLAG-peptide
TR-TRAP Complex

6. 0.25% Trypsin, 1 m*M* EDTA solution (Life Technologies, cat. no. 25200-056).
7. Phosphate-buffered saline (PBS): 0.14 *M* NaCl, 1.5 m*M* KH$_2$PO$_4$, 2.7 m*M* KCL, and 8.1 m*M* Na$_2$HPO$_4$ at pH 7.2.
8. Fetal bovine serum (FBS) (Life Technologies, cat. no. 10437-028).
9. Dialyzed FBS (10,000 MW cut-off) (Life Technologies, cat. no. 26300-053).
10. 100X Penicillin (5000 IU/mL) and streptomycin (5000 µg/mL) stock solution (Life Technologies, cat. no. 15070-063).
11. 100-mm Falcon tissue culture dishes (Becton Dickinson, Franklin Lakes, NJ, cat. no. 353003).
12. 6- and 24-Well tissue culture plates (Costar, Corning, NY).
13. Sterile syringe filters: 0.2 and 0.45 µm (Nalgene, Rochester, NY).
14. 0.25 *M* CaCl$_2$ solution.
15. 2X HEPES-buffered saline (HBS) solution: 0.28 *M* NaCl, 0.05 *M* N-2-hydroxy-ethylpiperazine-N'-2-ethanesulfonic acid (HEPES), and 1.5 m*M* Na$_2$HPO$_4$ at pH 7.05.
16. Polybrene: 4 mg polybrene (Sigma) dissolved in 1 mL ddH$_2$O.
17. G418 (Geneticin) solution: 50 mg/mL (Life Technologies, cat. no. 10131-035).
18. 100-, 500-, 1000-, and 3000-mL spinner flasks (Bellco Glass, Inc., Vineland, NJ).
19. 1 *M* Dithiothreitol (DTT) (Sigma).

2.2. Preparation of Nuclear Extract

1. HeLa S3 cells, Namalwa cells (American Type Culture Collection), or HeLa-derived cells stably expressing f:TR (*see* **Subheading 3.1.**).
2. RPMI medium 1640 (Life Technologies, cat. no. 11875-093).
3. Glass Dounce homogenizer: 15 mL with type B pestle (Kontes, Vineland NJ).
4. 2-Mercaptoethanol (β-ME): 14.3 *M* (Sigma).
5. 0.2 *M* Phenylmethylsulfonyl fluoride (PMSF): 3.48 g PMSF (Sigma) in 100 mL ethanol.

Fig. 2. (*Opposite page*) FLAG epitope-tagging as a method for purifying TR-coregulatory protein complexes from mammalian cells. (**A** and **B**) Generation of a HeLa-derived cell line that stably expresses a FLAG-tagged human thyroid hormone receptor (f:TR). A full-length TR cDNA (in this case, human TRα1) is first inserted inframe into the pFLAG(s)-7 plasmid *(12)* creating pFLAG-TRα. The *FLAG-TRα* gene is then subcloned into pBABEneo *(13)*, which is a retroviral transfer vector confering G418-resistance. The pBABEneo-FLAG-TRα construct is subsequently transfected into ψCRIP cells (an amphotrophic viral packaging cell line) *(14)*. High titer recombinant virus is then used to infect HeLa cells followed by G418 selection. (**C**) Strategy for purification of TR-associated coregulatory proteins (e.g., TRAPs). f:TR-expressing cells are cultured in media supplemented with thyroid hormone (T3) and then harvested for nuclear extract preparation. Anti-FLAG chromatography (M2-affinity resin), followed by FLAG-peptide elution is used to immunopurify and elute the TR-TRAP complex.

6. Hypotonic buffer: 10 mM Tris-HCl (pH 7.9 at 4°C), 1.5 mM MgCl$_2$, 10 mM KCl, 0.2 mM PMSF, and 10 mM β-ME. Add PMSF and β-ME immediately before use (*see* **Note 2**).
7. Low-salt buffer: 20 mM Tris-HCl (pH 7.9 at 4°C), 25% glycerol, 1.5 mM MgCl$_2$, 0.2 mM EDTA, 20 mM KCl, 0.2 mM PMSF, and 10 mM β-ME. Add PMSF and β-ME immediately before use.
8. High-salt buffer: 20 mM Tris-HCl (pH 7.9 at 4°C), 25% glycerol, 1.5 mM MgCl$_2$, 0.2 mM EDTA, 1.2 M KCl, 0.2 mM PMSF, and 10 mM β-ME. Add PMSF and β-ME immediately before use.
9. BC100 buffer: 20 mM Tris-HCl (pH 7.9 at 4°C), 100 mM KCl, 20% glycerol, 0.2 mM EDTA, 0.2 mM PMSF, and 1 mM DTT. Add PMSF and DTT immediately before use.
10. BC300 buffer (same as BC100 buffer except it contains 300 mM KCl).
11. Conductivity meter (Fisher Scientific Products, Pittsburgh, PA, cat. no. 09-327-1).
12. Dialysis membrane (MW cut-off 12–14 kDa) (Spectrum, Houston, TX).

2.3. Purification of the TR-TRAP Coactivator Complex from Nuclear Extract

1. 1 mM Thyroid hormone (T3) stock solution: dissolve 6.5 mg 3,5,3'-triiodothyronine (Sigma, cat. no. T2877) in 10 mL 10 mM NaOH. Filter-sterilize the solution through a 0.2-μm syringe filter and store at 4°C. The solution should be protected from light and is stable for 1–2 mo.
2. Anti-FLAG M2-agarose Affinity Gel (Sigma, cat. no. A-1205). Anti-FLAG monoclonal antibodies coupled to agarose beads.
3. Nuclear extract prepared from T3-treated HeLa-derived cells stably expressing recombinant f:TR (**Subheading 3.2.**).
4. 2.5 M KCl.
5. Compact reaction columns (United States Biochemical, Cleveland, OH, cat. no. 13928).
6. Nonidet P-40 (NP40) (Sigma, cat. no. N-6507).
7. FLAG peptide: Asp-Tyr-Lys-Asp-Asp-Asp-Asp-Lys (25 mg/mL) (Bacham, San Carlos, CA).
8. Silver staining kit (ICN Biomedicals, Inc., Aurora, OH, cat. no. 800665).

2.4. Purification of Baculovirally Expressed FLAG-Tagged TR and RXR from Sf9 Cells

1. pVL1392-FLAG-TR and pVL1392-FLAG-RXR plasmid DNA (*see* **Note 3**).
2. Sf9 Insect cells (Invitrogen, Carlsbad, CA).
3. Grace's insect medium (Life Technologies, cat. no. 11605-102).
4. 10% Pluronic F-68 solution (Sigma, cat. no. P5556).
5. Gentamycin (10 mg/mL) (Life Technologies, cat. no. 15710-064).
6. BaculoGold-modified baculovirus DNA (Pharmingen, San Diego, CA, cat. no. 21100D).

7. Insectin-Plus cationic liposomes (Invitrogen, cat. no. K3695-01).
8. Coomassie blue stain: dissolve 0.25 g of Coomassie Brilliant Blue R250 (Sigma) in 90 mL of methanol:ddH$_2$O (1:1 v/v) and 10 mL glacial acetic acid.

2.5 In Vitro Transcription

1. Template DNA (*see* **Note 4**).
2. Namalwa or HeLa cell nuclear extract (10 mg/mL) (*see* **Subheading 3.2.**).
3. 0.5 *M* HEPES-KOH (pH 7.9).
4. 0.25 *M* MgCl$_2$.
5. G-less NTP-mixture: 1.25 m*M* 3'-O-methyl-GTP, 6.25 m*M* UTP, 6.25 m*M* ATP, and 0.25 m*M* CTP. Stock nucleotide solutions are available commercially (Amersham Pharmacia, Piscataway, NJ).
6. [α-^{32}P]CTP (specific activity, 650 Ci/mmol) (Amersham Pharmacia).
7. 0.4 *M* DTT.
8. RNase ribonuclease inhibitor: 40 U/μL (Stratagene, La Jolla, CA, cat. no. 300151). Inhibits RNase A, RNase B, and RNase C, but not RNase T1.
9. RNase T1: 1 U/μL (Amersham Pharmacia, cat. no. 27-0991).
10. 200 m*M* KCL.
11. Carrier tRNA: 10 μg/μL (Life Technologies, cat. no. 16051-039). Dissolve lyophilized RNA in autoclaved ddH$_2$O at the appropriate concentration and store at –20°C.
12. Transcription-stop buffer: 7 *M* urea, 10 m*M* Tris-HCl (pH 7.8), 10 m*M* EDTA, 0.5% sodium dodecyl sulfate (SDS), 100 m*M* LiCl, and 300 m*M* CH$_3$COONa.
13. Phenol:Chloroform:isoamyl alcohol (25:24:1, v/v) (Life Technologies, cat. no. 15593-049).
14. Isopropanol (J. T. Baker, Phillipsburg, NJ).
15. DNA-loading buffer: 50 m*M* Tris-HCl (pH 7.9), 10 m*M* EDTA, 0.025% bromophenol blue, 0.025% xylene cyanol, 80% redistilled formamide (Life Technologies, cat. no. 15515-026).
16. X-OMAT-AR autoradiography imaging film (Eastman Kodak Company, Rochester, NY, cat. no. 165 1454).

3. Methods

3.1. Generation of a Stable Epitope-Tagged TR-Expressing HeLa Cell Line

HeLa cells and the retroviral packaging cell line ψCRIP (*14*) are routinely maintained in DMEM, 10% FBS, 50 IU/mL penicillin and 50 μg/mL streptomycin. The pBABEneo-FLAG-TR plasmid DNA (*see* **Note 1**) is first transfected into ψCRIP cells using the calcium phosphate-DNA precipitation method (*17*). Recombinant amphotropic viral particles are then collected and used to infect HeLa cells. Individual f:TR-expressing subclones are subsequently identified and expanded using G418 selection.

3.1.1. Preparation of Recombinant Retrovirus

1. In a sterile 4-mL culture tube, combine 20 µg pBABEneo-FLAG-TR with 0.5 mL HBS. Add 0.5 mL 0.25 *M* CaCl$_2$ solution drop-wise while gently vortex mixing tube. Let the tube stand at room temperature for 30 min.
2. Vortex mix the DNA/HBS/CaCl$_2$ solution vigorously for 1 to 2 s and then add it drop-wise to a 100-mm plate containing ψCRIP cells at 50% confluency in 10 mL DMEM/10% FBS. Gently swirl the plate to ensure uniform distribution of the DNA precipitate across the monolayer of cells. Incubate the plate at 37°C (5% CO$_2$) for 4 h.
3. Remove the media and glycerol shock the cells with 3 mL of 15% glycerol in DMEM (no serum). Let the plate stand at room temperature for 3 min. Aspirate the glycerol/DMEM solution and add 12 mL DMEM (no serum); gently swirl the plate and then remove media. Repeat the DMEM wash 3×. Finally, add 10 mL DMEM/10% FBS, and incubate the plate at 37°C (5% CO$_2$) for 18 h. Recombinant retroviral particles are released into the culture media.
4. Harvest the retrovirus particles by passing the culture supernatant through a 0.45-µm cellulose acetate syringe filter. Collect the filtered supernatant in a sterile 15-mL tube with a screw cap and store at 4°C.

3.1.2. Establishment of Stable f:TR-Expressing HeLa Cells

1. Aspirate the media from a 100-mm plate of HeLa cells at 50% confluency and add 2 mL DMEM/10% FBS containing 16 µg/mL polybrene followed by 2 mL of recombinant retrovirus stock (*see* **Subheading 3.1.1.**, **step 4.**); gently swirl plate to mix. Incubate the plate at 37°C (5% CO$_2$) for 2.5 h (*see* **Note 5**).
2. Add an additional 8 mL of DMEM/10% FBS to the plate and continue to incubate at 37°C (5% CO$_2$) for 36 h.
3. Remove the media and split the cells 1:10 with Trypsin-EDTA (e.g., the original 100-mm plate of retrovirus-infected HeLa cells should be split into 10 100-mm plates). To each plate, add 10 mL DMEM/10% FBS containing 1 mg/mL G418 (the active component is approx 0.5 mg/mL). The media should be replaced every 3 d (*see* **Note 6**). G418-resistant colonies are visible after 10–15 d. Individual colonies should be isolated when colony diameter is about 2 to 3 mm.
4. To select and clonally expand G418-resistant colonies, first pipet 100 µL of stock Trypsin-EDTA solution into each well of a 24-well tissue culture plate. Using the narrow end of a sterile 200-µL pipet tip, pick a clump of cells from the center of a well-isolated colony and then carefully place the wide end of the tip onto a micropipetor (*see* **Note 7**). Inoculate the colony into one of the wells of the 24-well plate by pipeting the cells up and down in the Trypsin-EDTA solution, thereby breaking the cell clump apart. Repeat procedure until each well contains cells from a different colony. Add 1 mL of DMEM/10%FBS/G418 to each well and continue to incubate at 37°C (5% CO$_2$). Replace the selective growth media with fresh media every 3 d.
5. When the cell monolayer in each well becomes confluent, detach the cells with 250 µL of the Trypsin-EDTA stock solution and transfer them to 6-well

plates. Add 3 mL of DMEM/10%FBS/G418 to each well and continue to incubate at 37°C (5% CO_2) changing the media every 3 d. When cells are confluent, whole cell lysate can be prepared and f:TR expression verified by Western blot, using antibodies against either TR or the FLAG-epitope (for details *see* **ref. *18***). Once identified, f:TR-expressing sublines can be maintained and propagated in 100-mm plates containing 10 mL DMEM/10% FBS/G418.

6. To adapt a f:TR-expressing subline into a suspension culture, the line should be expanded into 10 100-mm plates. Aspirate the media and add 2 mL of Trypsin-EDTA solution to each plate. Dislodge the adherent cell monolayer by pipeting up and down, and collect the detached cells (from all 10 plates) in a sterile 50-mL centrifuge tube. Add PBS (approx 30 mL) until the final volume reaches 50 mL. Gently mix by hand, and then pellet the cells at 250*g* (1200 rpms) for 5 min at 4°C. Aspirate the PBS and resuspend the cell pellet in 50-mL Jokliks medium containing 10% FBS, 50 IU/mL penicillin and 50 µg/mL streptomycin (*see* **Note 8**). Transfer the 50-mL cell suspension into a 100-mL spinner flask and incubate the flask on a stirplate at 37°C; the magnetic stir bar in the spinner flask should be completely submerged below the cell suspension. For quantitative cell harvesting (*see* **Subheading 3.2.**, **step 1**), suspension cultures should be gradually expanded into larger spinner flasks (500 mL, 1000 mL, and finally 3000 mL). Suspension cultures should be maintained at a density of 2–5×10^5 cells/mL.

3.2. Preparation of Nuclear Extract

This section describes a method for preparing nuclear extract from suspension cultures of HeLa, Namalwa, or HeLa-derived cells stably expressing f:TR (*see* **Subheading 3.1.2.**). The protocol is essentially based upon the method of Dignam *(19)* with a few modifications. Both HeLa S3 and HeLa-derived sublines are grown in spinner flasks containing Jokliks medium, 10% FBS, 50 IU/mL penicillin and 50 µg/mL streptomycin and maintained at a density of 2–5×10^5 cells/mL. Namalwa cells are grown in spinner flasks containing RPMI 1640 medium, 10%FBS, 50 IU/mL penicillin and 50 µg/mL streptomycin and maintained at a density of 1–2×10^6 cells/mL.

1. Six L of either HeLa or Namalwa cells in suspension culutre (e.g., two 3000-mL spinner flasks) are grown to density of either 5×10^5 or 2×10^6 cells/mL, respectively. Centrifuge the cells in sterile 1-L plastic bottles (approx 800 mL of culture per bottle) at 3000 rpm (1850*g*) for 15 min in a refrigerated RT7 centrifuge (Sorvall, Newton, CT) at 4°C with the brake turned off. Carefully pour off most of the supernatant and add fresh suspension culture on top of the existing cell pellet and repeat. When all of the culture has been centrifuged, pour off the supernatant until about 50–100 mL of media remains above the pellet. Gently resuspend the cells in the remaining media and transfer into conical 250-mL centrifuge bottles (Corning Costar Corp., Cambridge, MA). Centrifuge cells at 1850*g* (3000 rpm) for 15 min at 4°C with the brake on.

2. All subsequent steps are performed on ice or in a 4°C cold room using prechilled buffers and pipets. Aspirate media and resuspend cell pellet in ice-cold PBS until final cell suspension volume reaches 50 mL. Transfer cell suspension to a prechilled calibrated 50-mL centrifuge tube and recentrifuge as above.

3. Using the graduations on the tube, measure the packed cell volume (pcv) (*see* **Note 9**). With the tube on ice, resuspend the cell pellet in a volume of ice-cold hypotonic buffer approx 4 times the pcv. Centrifuge at 1250*g* for 10 min at 4°C. Carefully remove supernatant and resuspend cell pellet in a volume of hypotonic buffer approx 2 times the pcv. Allow the cells to swell on ice for 10 min.

4. Transfer cell suspension to a Dounce homogenizer on ice. Homogenize with 10–12 up-and-down strokes using a type B pestle. Verify lysis of outer membranes (*see* **Note 10**).

5. Transfer the homogenate to a calibrated 50 mL centrifuge tube on ice. Pellet the nuclei at 3000*g* (4000 rpm) for 15 min at 4°C. Carefully remove supernatant and measure the nuclear pellet volume (npv) using the graduations on the tube. Resuspend the nuclear pellet in a volume of ice-cold low-salt buffer equal to 1/2 the volume of the npv.

6. Transfer the nuclei suspension to the Dounce homogenizer on ice and homogenize with six up-and-down strokes using a type B pestle.

7. In a 4°C cold room, transfer the homogenate to a prechilled 10-mL glass beaker containing 1/4-inch magnetic stir bar. While mixing the homogenate gently on a magnetic stirrer, add (in a drop-wise fashion) a volume of high-salt buffer equal to 1/2 the volume of the npv. Continue to stir for 30 min in the cold room.

8. Transfer the extracted nuclei suspension to a prechilled polycarbonate centrifuge tube (Nalgene, cat. no. 3431-2526). Using a prechilled 50.2 Ti rotor (Beckman Coulter, Fullerton, CA), centrifuge the extract at 41,000*g* (18,000 rpm) in a XL-90 ultracentrifuge (Beckman Coulter) at 4°C for 30 min.

9. Transfer the supernatant into dialysis tubing (prerinsed in cold BC100) and dialyze against 1 L cold BC100 buffer for 3–6 h at 4°C changing the buffer several times. Dialyze until the conductivity of the nuclear extract equals that of the BC100 buffer (approx 60 μ*M* Ho). The protein concentration of the nuclear extract prepared in this manner is typically 10 mg/mL.

10. Transfer the dialyzed nuclear extract to a prechilled centrifuge tube and flash-freeze in liquid nitrogen. Store the extract at –80°C.

3.3. Purification of the TR-TRAP Coactivator Complex from Nuclear Extract

For experiments involving thyroid hormone (T3) induction of the f:TR-TRAP complex, stable f:TR-expressing HeLa cell lines (*see* **Subheading 3.1.2.**) should be expanded in Jokliks medium depleted of T3 for 3–6 d prior to hormone induction. Toward this end, we routinely expand the cells in dialyzed fetal serum (10,000 MW cut-off) (*see* **Subheading 2.1.**). Alternatively, one can deplete T3 from FBS manually by treating the serum with activated charcoal (Sigma) and the anion exchange resin AG1-X8 (Bio-Rad, Hercules, CA) *(5)*.

1. Expand the f:TR-expressing HeLa-derived subline (**Subheading 3.1.2.**) into two 3000-mL spinner flasks, each containing 1.5 L Jokliks medium, 10% dialyzed FBS, 50 IU/mL penicillin and 50 µg/mL streptomycin per flask. When each culture reaches a density of 5×10^5 cells/mL, add 1.5 L of freshly prepared medium containing 2×10^{-7} M T3 to each flask (giving a final T3 concentration of 10^{-7} M) (*see* **Note 11**). Continue to grow the T3-induced cells for 48–60 h or until cell density reaches 5×10^5 cells/mL.

2. Harvest cells and prepare nuclear extract essentially as described in **Subheading 3.2.** Six L of cells (at 5×10^5 cells/mL) should yield about 5 mL of nuclear extract.

3. Thaw the nuclear extract (if frozen) on ice in a 4°C cold room. Adjust the KCl concentration in the nuclear extract to 300 mM using an appropriate amount of ice-cold 2.5 M KCl. Verify the KCl concentration in the extract using a conductivity meter; the conductivity should be roughly equivalent to that of the BC300 buffer.

4. In a 4°C cold room, mix 250 µL (packed) pre-equilibrated M2 Affinity resin (*see* **Note 12**) with approx 5 mL of nuclear extract in a prechilled 5-mL culture tube (i.e., use 50 µL packed M2 resin per 1 mL nuclear extract). Securely seal the tube with parafilm and mix for 6–12 h (i.e., overnight) at 4°C on a rotator (one that rotates 360°).

5. Gently pellet the M2 resin in the cold room at 175g (1500 rpm) for 5 min using a swinging bucket rotor in a Centra CL2 clinical centrifuge (International Equipment Company, Boston, MA). Carefully remove the supernatant until approx 0.5–1 mL of supernatant remains above the pelleted M2 resin. Using a prechilled sterile transfer pipet, resuspend the M2 resin in the remaining nuclear extract and transfer the slurry into a sterile prechilled compact reaction column (CRC) containing a filter-plug but lacking a stopper. With the CRC tube attached to an upright support stand, let the nuclear extract drain out of tube by gravity allowing the M2 resin to pack inside the CRC tube (*see* **Note 13**). Insert a plastic stopper into the bottom of the CRC tube. The M2 resin should appear wet or moist. Do not let the resin dry out.

6. Attach a 5-mL syringe (minus plunger) to the top of the CRC tube using a leur-lock lid (supplied with the CRC tubes). Remove the plastic stopper and wash the packed M2 resin column with 5 mL prechilled BC300 buffer containing 0.1% NP40. Allow the buffer to drain out of the CRC tube by gravity and then repeat with a fresh 5 mL of BC300/0.1% NP40 (e.g., the M2 resin should be washed with approx 40 packed column volume of BC300/0.1% NP40). Finally, wash the column with 5 mL (approx 20 packed column volume) of freshly prepared BC100 buffer. Allow the buffer to completely drain out of the CRC tube and then insert a plastic stopper into the bottom of the tube.

7. To elute the f:TR-TRAP complex, add 250 µL elution buffer (BC100 buffer containing 0.2 mg/mL FLAG peptide) to the packed M2 resin (*see* **Note 14**). Securely attach a screw cap to the top of the CRC tube and mix the resin on a 360° rotator in the cold room for 1 h. Remove the plastic stopper from the bottom of CRC tube and (with cap partly unscrewed) place the CRC tube inside a prechilled microfuge tube. Collect the eluate by spinning at 175g (1500 rpm) for 10 s. Repeat the elu-

tion 1 or 2 more times (as above) by adding fresh BC100/FLAG peptide buffer to the M2 resin. Aliquot the eluted proteins in prechilled 0.5-mL microfuge tubes (approx 20 µL per tube) and then flash-freeze in liquid nitrogen. Store the extract at –80°C. The protein concentration of the TRAP complex (i.e., f:TR plus the TRAP subunits) is typically 20–40 ng/µL (with f:TR comprising about 10% of total protein). The f:TR-TRAP complex can be visualized by fractionating the proteins using sodium dodecyl sulfate-polyacrylamide gel electrophoresis (SDS-PAGE) followed by silver staining (**Fig. 3A**) (*see* **Note 15**).

3.4. Purification of Baculovirally-Expressed FLAG-Tagged TR and RXR from Sf9 Cells

Sf9 cells are routinely maintained at 27°C in tissue culture plates containing Graces medium, 10% FBS, and 10 µg/mL gentamycin. For growth of Sf9 cells in suspension cultures, 0.1% pluronic F-68 is added to the media. Initially, Sf9 cell monolayers are cotransfected with modified baculovirus genomic DNA (Pharmingen, cat. no. 21100D) and either the pVL1392-FLAG-TR or -RXR baculovirus transfer vector construct (*see* **Note 3**). The transfection supernatant (containing recombinant f:TR- or f:RXR-expressing baculovirus) is then amplified to a high titer and used to infect Sf9 cells growing in spinner flasks.

Fig. 3. Immunopurified FLAG-tagged nuclear receptors plus associated cofactors can activate transcription in vitro. (**A**) f:TR-TRAP complex preparation immunopurified from T3-treated HeLa cells stably expressing a f:TRα transgene (α-2 cells; *see* **ref. 5**). Immunopurified f:TR-TRAP complex (200 ng)(containing approx 20 ng f:TRα) was fractionated on an 8% SDS-polyacrylamide gel and silver-stained. TRAP subunits are indicated on the right. Subunits shown in parentheses indicate suspected human homologs of yeast mediator proteins designated as such on the basis of their size. Lower molecular weight TRAP subunits (<30 kDa) were too small to be resolved on this gel. TR-TRAP preparations containing additional subunits and purified under apparently modified conditions have been reported elsewhere *(8)* (*see* **Note 15**). (**B**) Baculovirus-expressed FLAG-TRα and -RXRα (bv-f:TRα and bv-f:RXRα, respectively) immunopurified from infected Sf9 cells. Immunopurified bv-f:TRα (100 ng) and bv-f:RXRα (50 ng) were fractionated on an 8% SDS-polyacrylamide gel and stained with Coomassie brilliant blue. (**C**) Immunopurified bv-f:TR, bv-f:RXR, and the f:TR-TRAP coactivator complex activate transcription in vitro. bv-f:TRα (10 ng) (lanes 1 and 2) or 100 ng of the f:TR-TRAP complex (containing 10 ng f:TRα) (lanes 3 and 4) were added as indicated along with 20 ng of bv-RXRα to a Namalawa cell nuclear extract (40 µg), and transcription was measured from the TRE_3 Δ53 template (50 ng). The reference template ML200 (25 ng), which lacks TREs, was added to each reaction as an internal control. T3 (10^{-7} *M*) was added to the reaction in lane 2. To confirm the involvement of the TRAP coactivator complex in mediating transcriptional activation, 300 ng of affinity-purified polyclonal antibodies against one of the TRAP subunits (α-TRAP220) *(20)* was added to the reaction in lane 4.

A
TR-TRAP Complex
TRAP 240
TRAP 230
TRAP 220
TRAP 170
TRAP 150
TRAP 100
TRAP 97
TRAP 95
TRAP 93
TRAP 80
FLAG-TRα
(p36)
(hMED7)
(hSRB11)
(hMED6)
205 kD
116 kD
80 kD
49 kD
Silver stain
B
bv-f:TRα
bv-f:RXRα
97 kD
66 kD
45 kD
31 kD
Coomassie Blue stain
C
T3
α-TRAP220
f:TR-TRAPs
bv-f:TRα
bv-f:RXRα
TRE3Δ53
ML200
1 2 3 4

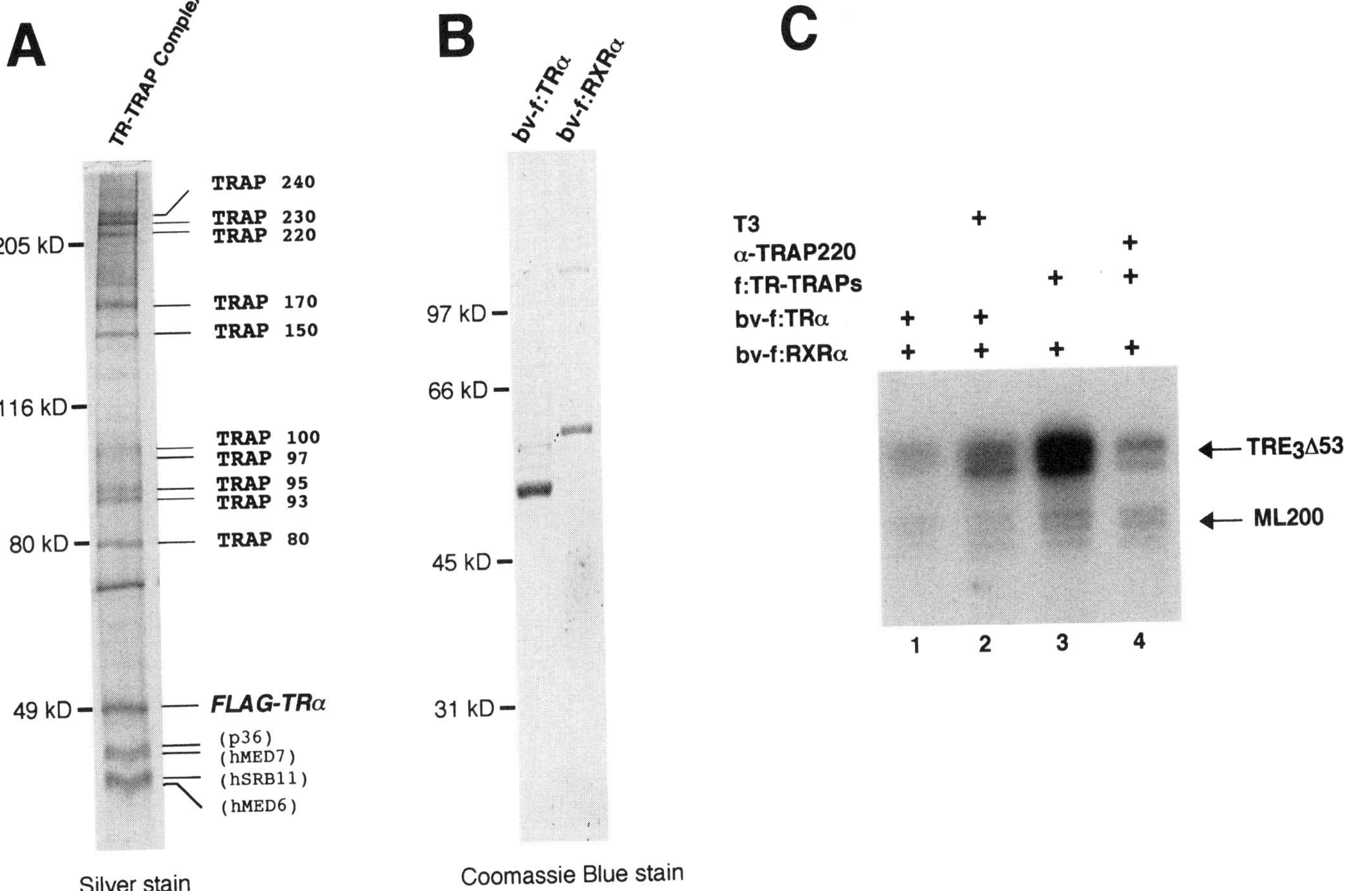

207

Infected cells are then harvested, nuclear extract prepared, and the recombinantly expressed f:TR- or f:RXR proteins are immunopurified using M2 resin.

3.4.1. Preparation of Recombinant Baculovirus

1. Seed 1×10^6 Sf9 cells into one well of a 6-well plate and incubate at room temperature for 1 h allowing cells to attach to the bottom of the plate. Aspirate media in the well and replace with 3 mL Graces medium (no serum). Allow cells to stand at room temperature while DNA is prepared for transfection.
2. In a sterile 1.5-mL microfuge tube, combine: 0.5 mL Graces medium (no serum), 250 ng BaculoGold linearized baculovirus DNA, 1 µg of either pVL1392-FLAG-TR or pVL1392-FLAG-RXR plasmid DNA, and 10 µL cationic liposomes (Invitrogen, Cat. no. K3695-01). Vortex mix vigorously for 10 s and then incubate at room temperature for 15 min.
3. Aspirate the media from the well (from **step 1**) and add the approx 0.5-mL transfection mixture (from **step 2**) to the cell monolayer in a drop-wise manner covering the entire bottom surface of the well. Rock the plate at room temperature on a platform rocker for 4 h. Adjust the speed to approx 2 side-to-side motions/minute.
4. Add 1.5 mL complete Graces medium (i.e., containing 10% FBS and 10 µg/mL gentamycin) to the well. As a negative control, seed 1×10^6 Sf9 cells into an empty well with 1.5 mL complete medium. Place the plate inside a plastic bag containing moist paper towels (to ensure humidity) and incubate at 27°C for 3–6 d.
5. Check the cells 3 to 4 d posttransfection for visual confirmation of transfection (*see* **Note 16**). If cells show no signs of viral infection, continue to incubate at 27°C for an additional 2 to 3 d and then recheck. When viral infection has been verified, harvest the cultured supernatant containing recombinant baculovirus particles. An aliquot of the transfection supernatant (approx 0.5 mL) can be stored at –80°C for future use. The remainder (approx 1 mL) can be stored at 4°C for 3–6 mo and should be used to prepare a high titer baculovirus stock as described below.
6. To amplify the titer of the recombinant baculovirus, combine 0.5 mL of the transfection supernatant (**step 5**) with 0.5×10^6 Sf9 cells in 1 well of a 6-well plate containing 3.5 mL (final volume) complete Graces medium. Rock the plate at room temperature on a platform rocker for 1 h and then incubate at 27°C for 4 d. After this time, the cells should show obvious signs of infection (>70% should be noticeably swollen and detached). Collect the cultured media supernatant (approx 3 mL) in a sterile 15-mL plastic screw cap tube and store at 4°C. This is the P1 titer.
7. In a 500-mL spinner flask, add 2 mL of the P1 baculovirus stock (**step 6**) to 5×10^7 Sf9 cells in 150 mL complete Graces medium containing 0.1% pluronic F-68. Grow the cells in the spinner flask for 7 d at 27°C or until most or all of the cells are dead. Transfer the cultured media into 50-mL centrifuge tubes and gently pellet

the dead cells and debris. Decant the supernatant into sterile 50-mL plastic screw cap tubes and store at 4°C. This is the P2 titer. Aliquots can frozen and stored at –80°C.

8. To prepare a large stock of high titer baculovirus, add 10 mL of the P2 titer (**step 7**) to 2.5×10^8 Sf9 cells in a 3000-mL spinner flask containing 500 mL complete Graces medium/0.1% pluronic F-68. Continue to spin the suspension culture until all of the cells are dead (about 7–10 d). Decant the cultured media into sterile plastic centifuge bottles and pellet the dead cells and debris. Decant the supernatant into sterile screw cap bottles and store in the dark at 4°C (*see* **Note 17**). This is the high titer baculovirus stock.

3.4.2. Expression and Purification of FLAG-Tagged TR and RXR in Sf9 Cells

1. In a 3000-mL spinner flask, grow Sf9 cells in 1 L complete Graces medium/0.1% pluronic F-68. When the cell density reaches 10^6 cells/mL, add 100 mL of the high titer recombinant baculovirus stock (*see* **Subheading 3.4.1.**, **step 8**). Continue to grow cells in suspension for 48–60 h (*see* **Note 18**).
2. Harvest the cells and prepare nuclear extract essentially as described in **Subheading 3.2.** with the following change: at **step 4**, homogenize the swollen cells with 20 strokes of the type B pestle in the 15 mL glass Douncer.
3. Immunopurify the FLAG-tagged proteins from the Sf9 cell nuclear extract as outlined in **Subheading 3.3.**, **steps 3–7**. The eluted proteins can be visualized by SDS-PAGE fractionation followed by Coomassie blue staining (**Fig. 3B**). The protein concentrations of the FLAG-tagged nuclear receptors purified and eluted, as described here, is typically 50–100 ng/μL.

3.5. In Vitro Transcription

This protocol describes an in vitro assay for measuring TR-dependent transcription using Namalwa cell nuclear extract as a source of basal transcription factors and RNA polymerase II (*see* **Note 19**). The purified f:TR-TRAP complex (**Subheading 3.3.**), together with purified baculovirus-expressed f:RXR (**Subheading 3.4.**), is added to the Namalwa extract (**Subheading 3.2.**) and specific transcription is measured using the G-less cassette reporter assay (*see* **Note 4** and **ref. 15**). In the protocol outlined here, we utilize a G-less cassette reporter gene (TRE$_3$Δ53, **ref. 16**) containing 3 TR binding elements (TREs) inserted upstream of the adenovirus major late (AdML) minimal promoter region (–53 to +10). The TRE$_3$ Δ53 template generates a specific 290-base transcript. As an internal control, a G-less cassette reporter gene (ML200, **ref. 16**) lacking TREs and generating a smaller 200-base transcript is included in each reaction.

1. Prepare a master mixture (*see* **Note 20**) on ice by combining in a 1.5-mL microfuge tube:

   1.25 µL 500 mM HEPES-KOH (pH 7.9)
   0.75 µL 250 mM MgCl$_2$
   0.25 µL 400 mM DTT
   0.5 µL RNase inhibitor
   0.5 µL RNase T1
   2.0 µL ddH$_2$O
   1.25 µL 200 mM KCl
   2.0 µL G-less NTP-mixture (*see* **Subheading 2.5.**)
   1.0 µL DNA Templates (50 ng/µL TRE$_3$Δ53 and 25 ng/µL ML200)
   0.5 µL [α-^{32}P]CTP (650 Ci/mmol)
   10.0 µL Total

2. In a separate tube, mix 4 µL of Namalwa nuclear extract (approx 10 mg/mL; **Subheading 3.2.**), 1–5 µL eluted f:TR-TRAP complex (20–40 ng/µL; **Subheading 3.3.**) and 1 µL baculovirus-expressed f:RXR (diluted to 20 ng/µL in BC100; **Subheading 3.4.2.**). Add BC100 buffer to bring the final volume to 15 µL (*see* **Note 21**). Incubate the mixture at room temperature for 10 min. As a control (e.g., T3-dependent TR activation in the absence of the TRAPs), set up two parallel reactions exactly as described here except that 1–5 µL baculovirus-expressed f:TR (diluted to 2–4 ng/µL in BC100; **Subheading 3.4.2.**) is added in place of the TR-TRAP complex. Into one of these control reactions, add T3 (10^{-7} M final).

3. To initiate the in vitro transcription reactions (*see* **Note 22**), aliquot 10 µL of the master mixture (**step 1**; *see* **Note 20**) into each of the 15 µL nuclear receptor–nuclear extract mixtures (from **step 2**). The final reaction volume is 25 µL. Pipet up and down 5–7 times to mix. Incubate the reactions in a 30°C water bath for 1 h.

4. To terminate the reactions, add 200 µL of transcription-stop buffer (**Subheading 2.5.**) together with 200 µL phenol:chloroform:isoamyl alcohol (25:24:1, v/v) to each sample. Vortex mix the for 10 s and then centrifuge at 20,000g (15,000 rpm) at room temperature for 5 min. Transfer the supernatant to a fresh tube and add 2.5 µL carrier tRNA (10 µg/µL) along with 200 µL isopropanol. Gently mix and then incubate at –20°C for 30 min.

5. Precipitate the nucleic acid by centrifugation (20,000g for 15 min) at 4°C. Resuspend the pellet in 15 µL DNA-loading buffer. Denature the sample at 95°C for 3 min, briefly centrifuge, and then quench the tubes on ice.

6. To resolve the ^{32}P-labeled transcripts generated in each reaction, load the denatured samples on a 5% (w/v) polyacrylamide gel containing 6 M urea. Run the gel at 300 V in 1X Tris-borate-EDTA (TBE) running buffer until the bromphenol blue reaches the bottom of the gel. Soak the gel in ddH$_2$O:methanol:glacial acetic acid (45:45:10 v/v) for 30 min at room temperature to remove the urea. Dry the gel under vacuum on a heated gel-dryer for 30 min and then expose to autoradiography film at –80°C with the aid of intensifier screens. Exposure for 5–15 h is usually sufficient to detect specific G-less transcript signals (**Fig. 3C**).

4. Notes

1. A prerequisite for immunopurifying recombinant TR from cultured mammalian cells is that the TR transgene should be epitope-tagged at either the amino or

carboxy terminus. FLAG-tagging at the amino terminus will facilitate efficient immunopurification of the TR-TRAP complex without any apparent disruption of normal TR transcriptional activity *(5)*. Toward this end, the full-length TR cDNA (TRα or TRβ) can be cloned into the pFLAG(S)-7 or pFLAG(AS)-7 vectors *(12)* as described *(5)*. Alternatively, TR can be cloned into commercial FLAG expression vectors (Sigma, St. Louis, MO). The resulting FLAG-TR gene should ultimately be subcloned into the polylinker region of pBABEneo *(13)*, a retroviral transfer vector conferring G418-resistance in transfected cells.

2. DTT can precipitate from aqueous solutions (especially those stored at –20°C) over the long term and should, therefore, be added to buffers immediately before use. Similarly, PMSF may precipitate to some degree in aqueous solutions immediately upon addition. To avoid this situation, PMSF should be added to buffers in a drop-by-drop manner with vigorous stirring.

3. The purification of recombinant TR and RXR from Sf9 cell extracts is greatly facilitated by again FLAG-tagging the receptor cDNAs at the amino terminus with an appropriate FLAG expression vector *(5)*. The resulting full-length FLAG-TR and FLAG-RXR gene fusions should then be subcloned into the polylinker region of either baculovirus transfer vector pVL1392 or pVL1393 (Pharmingen, San Diego, CA).

4. To eliminate background transcription, which may arise nonspecifically from various regions along the template DNA, we recommend the use of G-less cassette reporter genes *(15)*. The G-less cassette encodes a transcript lacking guanine nucleotides and can, therefore, be efficiently transcribed in vitro in a system lacking GTP. To eliminate potential background due to contaminating GTP (e.g., in the nuclear extract), the RNA chain terminator 3'-*O*-methyl-GTP is added along with RnaseT1, an enzyme that specifically cleaves G-containing transcripts at GpN. To generate specific G-less reporter templates, a strong eukaryotic promoter (e.g., adenovirus-2 major late promoter) is typically inserted upstream of the G-less cassette *(15)*. Various T3-response elements (TREs) can then be inserted upstream of the viral promoter site (e.g., pTRE$_3$Δ53, *see* **ref.** *16*), thus rendering the templates under the transcriptional control of TR. A DNA template lacking TREs (e.g., pML200, *see* **ref.** *16*) should be including in the in vitro assay as a control for promoter specificity (*see* **Fig. 3C**)

5. To ensure adequate exposure of recombinant retrovirus to HeLa cell monolayer, rock the plate back and forth by hand for a few seconds every 20 min and then immediately return the plate to the 37°C CO$_2$ incubator.

6. When changing the selection media, extreme care should be taken not to disrupt or dislodge any potential G418-resistant colonies, which may be forming on the bottom of plate surface. Toward this end, add fresh media into the plate slowly and aim the flow of media against the side wall of the culture dish rather than spraying it against the bottom of plate.

7. It is not necessary to quantitatively scrape up the entire cell colony. Rather, pick cells from the middle of the colony in much the same manner one would pick a bacterial colony from an LB-agar plate.

8. It is not necessary that the suspension cell cultures contain G418; in fact the cells will grow more vigorously without it. However, it is recommended that once identified, the various f:TR-expressing lines should be maintained in 6-well plates under G418 selection (*see* **Subheading 3.1.2., step 5**). Furthermore, it is recommended that aliquots of each subline be frozen and stored in liquid N_2 prior to adaptation to suspension culture.

9. Six L of HeLa suspension cells at a density of 5×10^5 cells/mL should yield a pcv of approx 5 mL.

10. After cell homogenization, lysis of outer cell membrane can be verified by microscope. Combine 10 µL cell homogenate with 100 µL PBS and 2 µL 0.4% trypan blue (Sigma). The nuclei of lysed cells stains bright blue while unbroken cells remain unstained. After 10 strokes, cell lysis should be ›80% complete; if trypan blue staining reveals ‹70% lysis, continue homogenization with an additional 1–4 strokes.

11. HeLa-derived cells will grow slower in dialyzed FBS than they do in normal serum. The normal doubling time for HeLa-derived cells is 24–36 h; in dialyzed serum, the doubling time increases to 48–60 h. T3-induction is usually timed such that the cells are exposed to T3 for a period of one cell-doubling prior to harvesting.

12. Pre-equilibrate the M2 Affinity resin in BC300 buffer overnight at 4°C with rocking. Use 1 mL of BC300 per 50 µL packed resin. Just before use, gently pellet the resin and then carefully aspirate off the BC300 buffer.

13. The volume of the CRC tube/column is approx 0.75 mL. Hence, transfer the M2 resin–nuclear extract slurry into the CRC tube in aliquots of 0.5 mL or less; allow all of the nuclear extract to drain out of the bottom of the column. When most of the slurry has been transferred out of the 5-mL tube, add a fresh 1 mL of BC300 buffer. Rinse and/or resuspend any remaining M2 resin on the inside of tube. Using the same transfer pipet, transfer any remaining M2 resin/BC300 slurry into the CRC tube. Repeat BC300 rinse until no resin is visible on the inside walls of either the culture tube or transfer pipet.

14. The volume of elution buffer should equal that of the packed M2 resin.

15. The overall subunit composition of the f:TR-TRAP complex (as shown in **Fig. 3A**) is relatively stable and constant under binding and washing conditions of up to 300 m*M* KCl. The quantitative purification of additional TRAP components (as reported in **ref. 8**) apparently reflects modified binding/washing conditions *(8)* and further indicates the existence of smaller mediator-like subcomplexes variably or weakly associated with a larger core TR-TRAP complex *(11)*. It should be noted that the TRAP proteins shown in **Fig. 3A** are transcriptionally active in vitro (**Fig. 3C**) and closely match the pattern of DRIP polypeptides associated with ligand-bound VDR *(9)*, thus suggesting that specific TRAPs essential for NR-coactivation are present in the TR-TRAP complexes prepared as outline here.

16. Confirmation of a successful transfection can be accomplished by comparing the transfected cells with the untransfected control cells using an inverted microscope at 250–400× magnification. Early signs of viral infection include a marked

swelling of the cells (e.g., a 25–50% increase in cell diameter) and a noticeable increase in size of the cell nucleus relative to the total cell volume. Late signs of infection include detachment of cells from the plate (e.g., an increase in the number of "floaters"), cessation of cell growth, and cell lysis.

17. We have found that the baculovirus titer will decrease by 10 to 100-fold after 6 mo of storage at 4°C. We, therefore, suggest that large volume of high titer viral stocks be prepared just prior to their use in experiments involving large-scale baculoviral expression of recombinant proteins in Sf9 cells (*see* **Subheading 3.4.2.**) Storage of the viral stocks in the dark (e.g., wrapping the bottle in aluminum foil) at 4°C appears to aid in maintaining a relatively high titer over prolonged periods.

18. For optimal recombinant protein expression the ratio of viral particles to Sf9 cells (i.e., the multiplicity of infection [MOI]) should be between 5 to 10. High titer baculoviral stocks prepared as described (**Subheading 3.4.1.**) and added to Sf9 cells as recommended (**Subheading 3.4.2.**, **step 1**) should be within this MOI range. The length of infection may also effect expression. Infection for periods greater than 72 h can lead to significant cell death, resulting in loss of viable protein from harvested cells. Conversely, infection for shorter periods (24–48 h) may be insufficient optimal protein expression. In our hands, infection for a duration of about 48 h results in the expression and purification of high levels of FLAG-TR and -RXR proteins (**Fig. 3B**) *(5)*. Researchers may want to perform pilot timecourse experiments in order to empirically determine the length of baculovirus infection, which yields optimal protein expression and purification.

19. In general, HeLa cell nuclear extract can used in place of Namalwa cell nuclear extract for the cell-free transcription assay. However, we typically observe higher levels of T3-dependent transcriptional activation using Namalwa extract.

20. For n reactions, prepare one stock master mixture for (n + 1) samples containing all of the common ingredients in order to minimize pipeting errors. From this stock mixture, pipet 10-µL aliquots into each transcription reaction (**Subheading 3.5.**, **step 3**).

21. Each protein component of this mixture is in BC100 buffer. The final KCl concentration of this 15-µL mixture should be 100 m*M*.

22. The optimal KCl concentration for in vitro transcription in a nuclear extract should be in the range of 60–80 m*M (19)*. In the example described here (**Subheading 3.5.**), the final KCl concentration is 70 m*M*.

References

1. Lazar, M. A. (1993) Thyroid hormone receptors: multiple forms, multiple possibilities. *Endocr. Rev.* **14,** 184–193.
2. Glass, C. K. and Rosenfeld, M. G. (2000) The coregulator exchange in transcriptional functions of nuclear receptors. *Genes & Dev.* **14,** 121–141.
3. McKenna, N. J., Lanz, R. B., and O'Malley, B. W. (1999) Nuclear receptor coregulators: cellular and molecular biology. *Endocr. Rev.* **20,** 321–344.
4. Moras, D. and Gronemeyer, H. (1998) the nuclear receptor ligand-binding domain: structure and function. *Curr. Opin. Cell Biol.* **10,** 384–391.

5. Fondell, J. D., Ge, H., and Roeder, R. G. (1996) Ligand induction of a transcriptionally active thyroid hormone receptor coactivator complex. *Proc. Natl. Acad. Sci. USA* **93,** 8329–8333.

6. Fondell, J. D., Guermah, M., Malik, S., and Roeder, R. G. (1999) Thyroid hormone receptor-associated proteins and general positive cofactors mediate thyroid hormone receptor function in the absence of the TATA box-binding protein-associated factors of TFIID. *Proc. Natl. Acad. Sci. USA* **96,** 1959–1964.

7. Myers, L. C., Gustafsson, C. M., Bushnell, D. A., et al. (1998) The Med proteins of yeast and their function through the RNA polymerase II carboxy-terminal domain. *Genes Dev.* **12,** 45–54.

8. Ito, M., Yuan, C.-X., Malik, S., et al. (1999) Identity between TRAP and SMCC complexes indicates novel pathways for the function of nuclear receptors and diverse mammalian activators. *Mol. Cell* **3,** 361–370.

9. Rachez, C., Lemon, B.D., Suldan, Z., et al. (1999) Ligand-dependent transcription activation by nuclear receptors requires the DRIP complex. *Nature* **398,** 824–828.

10. Ren, Y., Behre, E., Ren, Z., Zhang, J., Wang, Q., and Fondell, J. D. (2000) Specific structural motifs determine TRAP220 interactions with nuclear hormone receptors. *Mol. Cell Biol.* **20,** 5433–5446.

11. Malik, S. and Roeder, R. G. (2000) Transcriptional regulation through mediator-like coactivators in yeast and metazoan cells. *Trends Biochem. Sci.* **25,** 277–283.

12. Chiang, C.-M. and Roeder, R. G. (1993) Expression and purification of general transcription factors by FLAG-tagging and peptide elution. *Peptide Res.* **6,** 62–64.

13. Morgenstern, J. P. and Land, H. (1990) Advanced mammalian gene transfer: high titre retroviral vectors with multiple drug selection markers and a complementary helper-free packaging cell line. *Nucl. Acids Res.* **18,** 3587–3596.

14. Danos, O. and Mulligan, R. C. (1988) Safe and efficient generation of recombinant retroviruses with amphotropic and ecotropic host ranges. *Proc. Natl. Acad. Sci. USA* **85,** 6460–6464.

15. Sawadogo, M. and Roeder, R. G. (1985) Factors involved in specific transcription by human RNA polymerase II: analysis by a rapid and quantitative *in vitro* assay. *Proc. Natl. Acad. Sci. USA* **85,** 6460–6464.

16. Fondell, J. D., Roy, A. L. and Roeder, R. G. (1993) Unliganded thyroid hormone receptor inhibits formation of a functional preinitiation complex: implications for active repression. *Genes Dev.* **7,** 1400–1410.

17. Parker, B. A., and Stark, G. R. (1979) Regulation of simian virus 40 transcription: sensitive analysis of the RNA species present early in infections by virus or viral DNA. *J. Virol.* **31,** 360–369.

18. Sharma, D. and Fondell, J. D. (2000) Temporal formation of distinct thyroid hormone receptor coactivator complexes in HeLa cells. *Mol. Endocrinol.* **14,** 2001–2009.

19. Dignam, J. D. (1990) Preparation of extracts from higher eukaryotes. *Methods Enzymol.* **182,** 194–202.

20. Zhang, J. and Fondell, J. D. (1999) Identification of mouse TRAP100: a transcriptional coregulatory factor for thyroid hormone and vitamin D receptors. *Mol. Endocrinol.* **13,** 947–955.

Index

From: *Methods in Molecular Biology, vol. 202, Thyroid Hormone: Methods and Protocols*
Edited by: A. Baniahmad © Humana Press Inc., Totowa, NJ